THE SHEEP BOOK

Ron Parker

The Sheep Book

A Handbook for the Modern Shepherd

Revised and Updated

FOREWORD BY GARRISON KEILLOR | SKETCHES BY RUTH HALVORSON

SWALLOW PRESS | OHIO UNIVERSITY PRESS

ATHENS

SWALLOW PRESS | OHIO UNIVERSITY PRESS,
ATHENS, OHIO 45701

Top cover photograph, by Juliet M. McKenna, originally used on the
Paradise Fibers' website at www.paradisefibers.com. Bottom cover photograph courtesy of Janet McNally.

Swallow Press | Ohio University Press books are printed on
acid-free paper ∞ ™

10 09 08 07 06 05 04 03   5 4 3 2

Library of Congress Cataloging-in-Publication Data
Parker, Ronald B.
         The sheep book : a handbook for the modern shepherd / Ron
Parker ; foreword by Garrison Keillor ; sketches by Ruth Halvorson.—
Rev. ed.
     p.   cm.
Includes bibliographical references (p.).
ISBN 0-8040-1032-3 (pbk. : alk. paper)
  1. Sheep.   I. Title.
SF375 .P37   2001
636.3 —DC21

2001017004

CONTENTS

FOREWORD

The life of a shepherd is a temptation I have resisted so far, and when I see sheep at a fair I suppress the urge to make them a part of my life, for which those sheep ought to feel mighty grateful. My pasture in the city would be no more than a light snack for a small flock, and my life is too comical for sheep and too fastidious. City life has bred in me an instinct to travel with herds while keeping my distance and avoiding eye contact, and the idea of grabbing someone and prying open his mouth or examining the hindquarters is very far from my mind.

Still, it's good to examine a life one doesn't choose and to see what was missed, which is why my son and I drove up to visit the Parkers in the summer of 1982. My journal for Monday, July 12, reads, in part:

Got to Ron and Teresa Parker's place at one. They have 93 acres on which they raise sheep bred especially for their fine long wool which the P's sell (by mail) to woolspinners and knitters. Moved here six years ago, lived in an Airstream trailer for six months with infant son, then moved into one-room workshop with solar panel roof temporarily while building large circular solar home nearby which they are still building.

RP: "I grew up in L.A., Teresa in Wichita, neither of us on a farm, so when we came out here we had never handled animals before, other than dogs and cats. At first we tried to drive the sheep from pasture to barn and barn to pasture—running at them and clapping our hands and barking like we imagined sheepdogs bark—and we just about ran ourselves silly. The sheep would move a little ways, turn around, and stare at us. Finally we discovered that sheep can be led. They're out there wandering around and looking for a leader. You walk up to them and at the right moment you turn and they will follow you, especially if you reward them with grain now and then. I say, 'Sheep sheep sheep' to call them, or if they're a ways off, I call 'o-VEE' which is slang Latin and which I read in Thomas Hardy's *Far from the Madding Crowd* and which works.

"Sheep are stoics. We perform operations on them that would put a human in bed for a week, but the sheep get right up and walk away from it. One of Joshua's little rams had kidney stones, so we opened up his belly, cut off a few inches of the urethra, and strung it out below his anus, which is a handy thing to know. He's up and walking.

"All the sheep have names, and the triplets we try to name in series, such as Good, Bad, and Indifferent, or Shirley, Goodness, and Mercy. We lamb in April and by now we're ready to ship some to market. We breed them for wool, which we sell for $4.00 or so a pound (ordinary wool goes for 50 cents/lb. or less), but still, wool accounts for only a quarter of our income; meat provides the rest. We slaughter a couple of sheep a year for ourselves, which is hard to do but you just steel yourself and do it."

The P's have a fine little sheep dog, Spot, black with white chest markings, who climbed all over us, since she had been chained up for two days while the P's were in Mpls. After lunch we baled hay with them, using an antique Minneapolis-Moline tractor and John Deere baler, then had supper (bok choy).

RP: "It's almost cheaper to buy hay than raise your own; in fact it is. You can get hay for $1.25/bale, which is close to what it costs me to bale, or less, but—I'm a farmer and I like to do it myself. Most of the farmers around here don't earn much. It'd be much smarter to invest money in government bonds than in a farm; you'd earn more—but we're farmers. That's what we like to do. So we do it."

RP is also a writer. Sold article on sheep to *Country Journal* and is writing a book about sheep-raising for Scribners. Has written some short fiction with a sheep farm setting which he sends to *Atlantic* and *NYer* and collects rejection slips.

Were invited to spend night in Airstream but declined due to high-way fever & also felt if we stayed one night we might want to stay for the summer and beyond. They are gentle & attentive people & good company, which I find true of most farmers. Wonder if, in all the technical writing on animal husbandry, anyone notes the civilizing influence animal raising may have on people. Maybe it's simply understood by all—that for all the modern discoveries, shepherding is an ancient scientific culture & teaches people more than they intended to learn & brings out qualities in them they might not attain directly through moral ambition.

I THOUGHT SO then and think about it now as I sit at home in the city, worn out after a day of trying to rush a piece of fiction to a state of perfection and beating it to death in the process. Nobody rushes sheep—to perfection or anywhere else. Perhaps if sheep were part of my life, they would impose an order on it and bring out in me the calm patience and good humor so evident in this book. Perhaps you, dear readers who raise sheep or are thinking about it (or raise sheep *and* are thinking about it), can take some pleasure in knowing someone envies you and your enterprise. It's a pleasure to envy you and imagine what sheep raising is like, and if I'm all wrong—well, no sheep will ever suffer as a result.

—GARRISON KEILLOR

PREFACE TO THE REVISED EDITION

Sheep haven't changed much since 1983, but many things about sheep stewardship have changed. There are new medications, and a few that are no longer available. Artificial insemination and embryo transplants are almost routine now. A pregnant ewe can be scanned with ultrasound to see how many lambs she is carrying, and starting with the now famous Dolly, sheep are even cloned. What's more, if you want to know about some genes in a sheep, don't guess from breeding records—get a DNA analysis.

On the negative side, some diseases have appeared in sheep that were formerly found only in other species, especially cattle, and many sheep diseases that were confined to one part of the country have spread along with the widespread transportation of sheep. Worms have not given up trying to make their living inside sheep, and careless use of anthelmintics has built resistance in worm populations. Many antibiotics and other chemicals have been removed from the market after discovery that they had become ineffective or were dangerous in some way.

That said, new breeds are appearing in North America. A milk sheep industry is growing slowly. New shepherds are learning their ways with sheep, and new flocks are being started in many areas where sheep have not been common in the recent past.

The Information Age has not left sheep and shepherding behind. All manner of sheep and sheep-raising information is available on the Internet. The sheep-L mailing list, physically located in a computer in Uppsala, Sweden, has hundreds of subscribers from all over the world who exchange information and chat in as many as a hundred or more email messages a day. One can get information about breeds of sheep that most people have never even heard of at www.ansi.okstate.edu/breeds/sheep/ and many sheep breed associations have web sites to tell how wonderful their breed is. Information on sheep health is widely available from authoritative online sources. Wool enthusiasts and fiber arts/crafts people can start at my web site at www.rbparker. com to find many links and references to mailing lists, web sites, and all manner of information. Fiber people can subscribe to my mailing list, FiberNet, celebrating its eleventh anniversary in 2001, by sending an email to fibernet-request@imagicomm.com with the word "subscribe" as the message, or by visiting my web site.

So, I hope you enjoy this version of *The Sheep Book.* If you have any comments, I can be reached by email at ron@rbparker.com.

I want to give hearty thanks to Alexandra Weikert of Munich, Germany, who generously scanned a copy of the original edition of *The Sheep Book* and converted the information into text for me to work with. Alexandra did that after several subscribers to my FiberNet mailing list volunteered to type the book into their computers by hand. What better testament could one have to the togetherness of sheep and fiber people than such unselfish willingness to help. My constant companion, Susanne Kallstenius, also devoted countless hours to working with data and graphics that became part of this version, without which I would never have met the deadlines.

—RON PARKER
Täby, Sweden

PREFACE TO THE FIRST EDITION

When I quit a perfectly good job as a tenured professor in one of the country's top geology departments to move to Minnesota with my wife and son to raise sheep, did I bring with me a background in farming and raising livestock? I did not. Was my wife from rural roots and full of knowledge of things pastoral? She was not. In fact, we didn't know a thing about sheep, or very little at least. My wife had taken up spinning and weaving, and we thought we should raise some animals, so sheep it was. I sat in on some courses in the agriculture college at my university, and we read what books we could lay hands on, but we were mighty ignorant of both theory and practice. Oh, we had seen sheep; we had even sheared a sheep apiece at a clinic for 4-H kids, but that was about it.

We hurriedly fenced a pasture and moved in our newly acquired flock of twelve wild, crazy ewe lambs. They prospered on the abundant grass and tasty water, and we even caught sight of them now and then as they cruised around their pasture like so many gazelles. As breeding season approached we knew that we would have to catch them and remove the yearling ram and one ram lamb. We tried in vain to lure them into a small corral. We even enlisted the aid of a pair of neighbors, but, under the leadership of a half-Karakul ram whose instincts had reverted to those of his ancestors in the wilds of Persia, they outmaneuvered us with disheartening grace.

We knew there had to be a better way, so we did what any sensible person would have done: built a trap. Not an elaborate trap, mind you, just an enclosure at a fence corner that looked innocent enough even to a suspicious sheep with a love of freedom. We baited our trap with some corn after giving the sheep enough free samples to get them hooked. They moved in with cautious hunger, and my wife, prone in the grass, quickly reeled in the fifty-foot rope that slid the carefully balanced and lubricated door shut behind them. At last, we had our flock in captivity.

We knew that setting traps was no way to become shepherds, so we started to watch the sheep to learn what made them tick. We kept them confined, and pretty soon had them rushing in for grain and hay. We also discovered that sheep don't want to be chased. They want to be led. We changed our strategy to my leading and my wife coming along behind, clapping, and now and then—partly as a joke, and partly because it worked—barking like a sheep dog.

Things have come a long way since that period seven years ago. The sheep follow me anywhere I call them, and we even have a sheep dog who rounds them up for us (without barking, I might add). No longer do we desperately call our retired shepherd neighbor when something goes wrong. It no longer takes us an hour to shear a sheep. People even phone and ask us for advice.

Over the past few years, as we talked with sheep raisers, referred to some books on sheep raising and to our *Merck Veterinary Manual,* called our vet or took animals to him so we could watch what he did to doctor a sheep, we wished that we had a modern, complete book on sheep raising. As we learned more, I began to write short articles about what we thought would be helpful to the beginners we had once been—first for *Countryside,* then later for *The New Farm, The Shepherd,* and *sheep!* I finally got brave enough to try a book, and with lots of encouragement from editors, especially Bill Allison, the first editor of *sheep!,* and from my always understanding and sustaining wife, I gave it a try. What you see is almost the book that I wish I had had when I first got into the sheep and shepherding business.

So, read the book, get yourself some sheep, find a good veterinarian, and go to it. The sheep will teach you if you give them half a chance. Go to meetings of sheep organizations, read magazines, books, newspapers, and learn. The more you know about sheep, the more you'll love and enjoy them.

I want to thank the many people whose brains I picked, although I won't mention their names because I don't think it's fair for them to take any of the blame for my somewhat opinionated handbook. I do want to thank my wife, Teresa, and our close friend Kathy Barber for the many hours they both spent reading the first draft of this book and for setting me in the right direction more than once.

—RON PARKER
Sammen Sheep Farm
Henning, Minnesota
1983

Introduction

Sheep are the ideal, useful domesticated animal. They are hardy and healthy. Except for an occasional aggressive ram or uppity ewe they are gentle and submissive. They are small enough for a good-sized child or a senior citizen to handle. They give both superlative meat and a fiber that has no peer. They are the ideal animal for the homestead, small farm, place in the country, suburban backyard, or any other place where man makes his home and grass will grow.

Sheep don't get a lot of press coverage; they are shyly in the background while horses, cattle, hogs, rabbits, and even llamas catch the eye of the public. Still, though their flashier confrères may bathe in the limelight, sheep emerge as the old reliables. The first food-producing animal ever domesticated by man quietly eats grass and leaves and goes about its business of producing lambs and wool.

Sheep have a quality that makes them almost a symbol of rural peace and tranquillity. Landscape painters know that a sylvan scene can be rendered really irenic only if a band of grazing sheep and perhaps a shepherd are included in the composition. And can anyone listen to Beethoven's Sixth or Copland's "Appalachian Spring" without visualizing a flock of sheep at some point? This quality of sheep is a result of their mutual dependency, their soft gregariousness, and their ready acceptance of a leader-shepherd. They are endearing because they respond to loving care so willingly, and even flighty, frightened ones follow the familiar herder with eager timidity. They thrive in man's care, and return abundance to those who will care for them. Sheep alone among domesticated species give back innocence and gentleness to almost any treatment, and they can elicit the best from the best of us, those who are willing to give of themselves unselfishly.

Sheep have been domesticated for so long that even the scholars who specialize in such matters disagree about the precise identity of their wild precursors. Sheep as we know them today have coexisted with man for so long that they are more a part of human culture than beasts of the field.

The popularity of this domestic animal par excellence is enjoying a well-deserved renaissance in America. Sheep numbers declined in the decades after

World War II because of the combined impact of the returning GIs' dislike of the mutton served to them in the Pacific Theater, the explosive growth of the synthetic fiber industry, an increase in production and promotion of fowl and swine, the conversion of small and diversified farms to a few large-scale, cash-crop production units, the loss of sheepherders to higher paying and less demanding employment, and the impact of reduced predator control in the West.

The sheep renaissance is taking place not on the vast rangelands of the West but on thousands of small farms and homesteads all over the United States and Canada, especially in the Pacific Northwest, the Midwest, and the Northeast. The expansion in numbers is a reflection of an increase in the number of flocks, from handfuls to hundreds or thousands. The average number of sheep per producer in Minnesota is only about thirty. The new sheep raiser might be a farmer but just as often is a doctor, a mechanic, a college professor, a trucker, an artist, a homesteader, a retired person, or a pastor. The modern shepherd is a person looking for a degree of self-sufficiency and an involvement in producing his own food and even clothing, or may be trying to make a living at shepherding. He or she is a person who wants input into life that is lacking when every need is provided ready-made by a complex and structured human society.

The sheep raiser of the 2000s is a person who wants to be involved with the sheep and wants to do the job of shepherd in a professional way. The modern shepherd reads magazines and books, goes to meetings and seminars, joins sheep raisers' organizations, subscribes to sheep-related mailing lists on the Internet, visits other sheep owners to learn from them, and strives to attain a level of competence undreamed of by the farmer of yesteryear who might have kept a few sheep around the farmstead to trim the brush down. Today's shepherd recognizes the benefits that a complex, technologically oriented society can confer on the agrarian way of life, and is likely to keep current on the latest drugs, techniques, and management practices to a degree that may surprise extension agents, veterinarians, and other animal husbandry professionals. The shepherd of the 2000s uses a computer, may synchronize breeding and lambing times with biologically active chemicals, and generally does things that would have been thought avant garde even at university experiment stations only decades ago. This shepherd is reversing the evolution of the shepherd David to a great leader by coming from a position of expertise in middle-class society to become guardian of a sheep flock. Yet, the modern shepherd brings along a sophistication that the young David needed many years to learn, and acquires humility and a new sense of personal worth with the sheep as David did when he was a lad.

Sheep are an ideal animal for the small grower. They thrive on care yet will utilize hilly, rocky land that other animals disdain. Given fencing to prevent their wandering away, and protection from predators, sheep will take care of themselves provided enough grazing, browsing, and water are available. A flock of sheep will turn a brushy, weedy pasture into a place that resembles a well-groomed park. Their wool coat gives them such protection from the elements that they seldom require shelter, even in conditions that would be fatal to other animals.

A ewe births and raises one or more cuddly lambs each year that grow up to be sturdy young sheep ready for market in four to six months. The ewe and her lambs provide a clipping of wool each year that can be sold or spun into yarn for garments. Indeed, many of today's small sheep flocks were started because the shepherd wanted some wool for spinning, and many a fiber artist finds that raising sheep is at least as interesting as spinning and weaving. Some of them become shepherds of large commercial flocks, as Teresa and I did.

The new shepherd soon realizes that sheep are not something to be dominated but finds instead that they have a wide range of capabilities that the shepherd must learn. A wise shepherd today adapts lifestyle and schedule to the biological timing and needs of his flock, just as the wise shepherds of ten millennia ago did. The observant herder adjusts his annual work cycle to the natural cycle of the ewe, one that is strongly seasonal compared to that of animals such as swine, cattle, or horses. The sage shepherd becomes the servant of his flock and alters her ways and those of the family to function in a mutually beneficial rhythm according to the ewe's cycle of estrus, gestation, and parturition.

It is essential to learn how sheep behave and do it their way. Treat a sheep like a sheep, and it will reward you by doing as it should. Don't try to make a sheep like a dog, or a child, or a horse. Instead, spend time learning about how sheep operate, and guide them in what they will do naturally. Learn that a sheep has a safety zone around it, and if you enter that zone, it will flee. Learn that sheep will follow you if you give them a chance.

In recognition of the shepherd's ability to adapt to the essentially unalterable dictates of the ewe's biological clock, I have organized the core of *The*

Sheep Book according to the major sexual cycles of the mature ewe. I begin with the period of building and rebuilding, when a young ewe grows to maturity, or when an older one rebuilds her body after lambing and lactation. The physically prepared ewes are made ready for their role as mothers by a period of high nutrition called flushing. This is followed by breeding, gestation, lambing, lactation, weaning, and finally the rebuilding of resources for the beginning of a new cycle.

This book is written for everyone who is a shepherd or would become a shepherd. The size of the flock is not a relevant issue. Regardless of whether the flock is a couple of ewes in the backyard, a few hundred on a farm, or ten thousand on the range, the needs of an individual ewe remain the same. Her world is dominated by her natural cycle and the shepherd's responses to her wants. It is likewise not relevant whether the sheep are raised for their wool, their lambs, or as pets, nor does it matter whether profit is a motive. A good shepherd must have a feeling for the essential biological and psychological requirements of the flock that comes from a blend of observation, intuition, study, and a love for shepherding.

The book should first be read through from start to finish in order to get a sense of the ewe's rhythm. The details of chores, facilities, and health matters are not vital the first time through. When you have read the whole book you will be ready to read some parts again, more carefully. Use the index to help you locate where the same question is treated in the context of different stages of the ewe's cycle. Compare your sheep to those described in the book. Ask yourself how they are the same and how they differ. Test your own perceptions against those detailed in the book. What have you missed, and, conversely, what have you seen that isn't mentioned in the book? Read it for pleasure, use it as a reference, and, most important, use it as a starting point from which to become a better shepherd. When you think you know it all, buy some more books, watch some more sheep, and deepen your knowledge.

If you have access to a computer and a connection to the world, join the sheep-L mailing list and learn from an informed, chatty, sometimes rowdy, almost always cheerful group based all over the world who share an interest in sheep. Send email to listserv@listserv.uu.se with the message "subscribe sheep-L" and you will start to learn about sheep the first time you get your list mail.

Although the book is geared toward sheep raisers in North America, not every generality or piece of advice will fit every single situation even there.

Some shepherds do certain things at a different stage in the cycle or at a different time of year than I have assumed. There are also discussions of some activities that are not tied to the ewe's reproductive cycle at all, which are noted where necessary. The cardinal rule for today's shepherd, as it was for yesterday's, remains: let your ewes be your guide.

1

BUILDING AND REBUILDING

There is a period before the breeding season when the flock has a quiet time. The lambs have been weaned and are thriving on solid food. They run and cavort in frenzied packs, then eat, then rest, then play again, as if thoroughly enjoying being lambs. The rams are together in their own pasture in a male-only group. The ewes are mostly pretty thin and worn out from the demands of motherhood, and they are eager to eat and put back the weight that was lost during lactation. The experienced shepherd admires the really skinny ones in the bunch because they are usually the ones that are the best milkers and raise the biggest, healthiest lambs. Ewes that come out of lactation in good flesh are viewed with suspicion. They weren't doing their job or they would be as thin as the ones that gave of themselves.

The flock will change before the rams are put back with the ewes. The best ewes will rebuild their bodies. Some of the older ones will be culled from the group. Some ewe lambs—the ones with the right ancestors, or the fast growers, the twins and triplets, and others who are special for some reason—will be saved to join the flock as replacements. They are the new members of the club who will be given a chance to prove themselves, and who may become permanent members. The flock is built and rebuilt in this way every year, evolving little by little toward a goal of perfection that exists in the shepherd's mind.

Some would-be sheep raisers will buy their first sheep at this time—usually summer—when older ewes are rebuilding and ewe lambs are growing to sexual maturity. Each shepherd will have different reasons and different goals.

BUYING SHEEP

Selecting a Breed

When I first thought about writing this book I told myself that one thing I would leave out was advice on buying sheep, since each buyer's needs are unique. I've since changed my mind, because I thought back to when we bought our first sheep and decided that we could have used some suggestions, even if we didn't follow them, simply because suggestions get one thinking.

Before buying sheep, consider why you want them at all. Do you want to raise fast-growing lambs for a fat lamb market, produce specialty wool for handspinners, or have a few sheep to trim the lawns and keep brush under control, for pets, for 4-H projects, or for show stock? If you are sufficiently organized to know what you want, then you are probably familiar with sheep to some extent. Otherwise, start looking at sheep in your area and talk to sheep raisers at fairs and on their farms. Write to the secretaries of the various breed associations (see appendix 6) for literature that describes the various breeds. Read about different breeds and crossbreeds in magazines and books. You will feel overwhelmed with information at first, but after a bit you'll begin to form your own ideas.

A few suggestions might help you decide on a breed. If you are going to show sheep, or if youngsters in the family want to do so, then you should visit shows and see what classes of sheep are shown in your area. For example, you wouldn't want to raise Lincolns or Cotswolds if there were no long-wooled class at local and state fairs. You wouldn't want black or colored sheep if a whites-only rule prevented them from competing. Talk to the winners and the judges to get their views. You'll find the winners only too eager to sell you some high-priced stock, but keep your wallet in your pocket until you have accumulated some knowledge. If you choose Suffolks, be sure to ask the breeder about "spider lambs," which are lambs with unusually long legs—who generally do not survive. The spider gene appeared in some show flocks when breeders bred for tallness. John Beever, USDA research fellow at the University of Illinois, developed a blood test for the spider gene.

If you want to raise specialty-wool sheep, talk with spinners, weavers, and other fiber artists, and learn how to spin so that you can understand the needs of a handspinner. Also, ask yourself if you are willing to do the marketing of such wools: There are no established channels for selling handspinning fleeces unless you can contract with a shop to take your entire production. Are you prepared to maintain a standard of wool cleanliness that is virtually impossible for the average wool producer but essential to the handspinners' market?

Perhaps you just want a few sheep around as pets and decorative lawn mowers. The Cheviot breed was supposedly developed to look attractive on the lawns of the queen's summer castle, Balmoral, in Scotland. You may agree with this royal taste. We have had people buy black sheep from us for pets just because they wanted something a bit unusual. If you don't want to bother with breeding and lambing, you might even consider getting a few attractive wethers to keep around. They usually have good dispositions and make fine pets.

If you are going to try to raise sheep for a profit, then your choice automatically becomes a little more limited. You then want a breed that produces a lamb that is acceptable to lamb buyers and packers so you can command a top market price. You also want a breed that gives lambs that reach market weight quickly on a minimum amount of feed. Not only that, you want a breed that produces plenty of lambs by either having lots of multiple births or by breeding more often than once a year, or both.

Good carcass traits are found in most major breeds, though you may find that your local buyers have strong preferences and prejudices. Some breeds, Finnsheep and Karakul, for example, may not have good carcass conformation. In contrast, Columbias, Hampshires, and Suffolks are common meat breeds in all parts of the country. Many shepherds select a type of ewe for a given set of traits and choose a ram of a meat or mutton breed to sire market lambs.

Fast growth is encouraged by a number of management factors, but genetics is also significant. Suffolks are the acknowledged champions of growth rate, with Columbias a close second. Other breeds generally trail somewhat, at least insofar as purebreds are concerned. Characteristics such as fast growth are usually traceable to individual sheep, and the selection of replacements on the basis of growth rate can produce crossbreeds that equal or exceed the purebreds. Even with crossbred sheep, however, some of the purebred characteristics come through. In a group of mutton-type crosses with one-quarter Finn blood, we found that the growthiness of the Suffolk and Oxford crosses was better than that of the Hampshire crosses when taken as a group. One could, of course, choose other individuals and reverse the order.

Considering only such factors as lambing percentage and rate of gain may lead to an incorrect conclusion as to which breeds are the most productive. What matters is the number of pounds of lamb that reaches the market from each ewe. A breed that gives lots of twins and triplets is valuable only if they all live to be marketed. In the same way, a fast-growing lamb is of value only if it survives to shipping weight. G. E. Dickerson and others reported in 1981 on a study made at the U.S. Meat Animal Research Center in Clay Center, Nebraska, that bears on this question. They compared lamb production from sires of the three blackfaced breeds: Suffolk, Hampshire, and Oxford. The lambs sired by Suffolks excelled in rate of growth and in boned cuts per lamb, as might have been expected. However, the survivability of Suffolk-sired lambs was low, and the Oxford sires actually produced more pounds of boneless cuts per ewe than either Suffolk or Hampshire. In another comparison, they looked at performance of crossbred ewes by breed of dam (all the ewes were half Finn and half some other breed). In terms of lambing percentage the Suffolk-cross ewes were tops, but in terms of lambs weaned the Dorset, Targhee, and Rambouillet crosses beat the Suffolk crosses, with Corriedale and Hampshire crosses bringing up the rear.

A comparison of ewe breeds made at the Colby, Kansas, Agriculture Experiment Station by Frank Schwulst in 1982 sheds some light on the Suffolk survivability question. Purebred Rambouillet, Rambouillet × Dorset, and Rambouillet × Suffolk ewes were bred to Suffolk sires. Table 1 summarizes the results for fall 1980 lamb crops.

If you think these overall lambing percentages are low, remember that these are fall lambs, born out of the regular lambing season. The lambs with the most Suffolk breeding were the ones with the lowest survivability. In spite of the higher average market weight of the lambs from the Suffolk-cross ewes, the better livability of those from the purebred Rambouillets gave them a full 25 percent advantage in terms of weight of lambs actually marketed.

Table 1   Survivability of Suffolk-Sired Lambs

Ewe breed	Rambouillet	Rambouillet × Dorset	Rambouillet × Suffolk
Lambs born weaned (%)	91	88	80
Lambs born marketed (%)	89	88	72
Avg. market weight (lb.)	107	108	110
Lbs. lamb mktd/ewe lambing	114	104	91

Because of studies such as these, and based on our own experience comparing lambs sired by Suffolk with those sired by Lincoln and Finn × Lincoln rams, we stopped using Suffolk as a sire breed. The lambs just don't survive as well. On the other hand, just to confuse things, we have a ewe who is three-quarters Suffolk × one-quarter Finn and strictly outstanding. She and her daughters, and even her granddaughters, are among the most productive ewes in our flock, which points up the importance of individual traits as a factor. The point that I want to make is that you shouldn't be influenced too much by advertising that emphasizes a single factor, such as rate of growth, because there is more to the story than that.

For multiple births the Finnish Landrace, or Finnsheep, reigns. The closely related Romanov even exceeds Finns in some settings, with the disadvantage that the wool has no commercial value. Purebred Finn and Romanov ewes produce litters of a half dozen and more, but their carcass and growth characteristics are mediocre at best. Crossbred ewes of about one-half to one-quarter Finn are a delightful compromise, however, offering the best qualities of both breeds. Such crosses will give lamb crops averaging 200 percent and up (in other words, there is an average of two or more lambs per ewe over the whole flock) and will still produce lambs that grow rapidly and provide a good meat-type carcass. Some sheep raisers criticize Finn crosses, and packers complain about poor carcass quality, but a grader from a large livestock cooperative didn't realize that some of our quarter-Finn lambs had any Finn in them. One sheep raiser comments that the people who criticize Finns are the ones who have never tried them. This is not to say that other breeds and crosses cannot compete with the Finn crosses in prolificacy, but they do so only after many generations of selection and culling. If you are fortunate enough to be able to buy sheep from such a high-production flock, that's fine, but if not, the introduction of some Finn blood is the quick and easy route to a higher percentage of lambs born per ewe.

Finn crosses have the added advantage of producing unusually vigorous lambs. Lambs with one-quarter-or-more Finn blood are on their feet and sucking within moments after birth, and seem to just get down to the business of being healthy and aggressive lambs without hesitation. They also reach sexual maturity very early, which is an important factor if one wishes to breed ewes as lambs. I should add that this can be a nuisance if you have uncastrated ram lambs mixed with ewe lambs, because some of the ewes will get bred at four to six months, so the producer has to get the rams out of the lamb flock early.

A different, and equally useful, route to prolificacy is the Booroola gene, which was discovered in a Merino flock in Australia. The Booroola gene can be bred into any sheep breed. With Finns and Romanovs the prolificacy is related to a whole groups of genes, so the effect is tied to the amount of Finn or Romanov heritage in a ewe. In contrast, the Booroola prolificacy is tied to a single pair of genes.

A sheep can have two copies of the Booroola gene (BB), one copy (B+), or none (++). On average a BB ewe will have approximately 1.5 lambs more than the non-Booroola breed; B+ ewes will have .8 to 1.1 lambs more than the non-Booroola breed. A flock of BB ewes would produce too many lambs to be practical in most settings, but a flock of B+ ewes would be almost ideal for a farm flock. To get that ideal flock one need only breed all ++ (no Booroola gene) ewes to a BB ram. Then all the offspring would be B+. What next, though?

Here's how Janet McNally manages her Tamarack Booroola Dorset flock in Minnesota. The B+ ewes are bred to purebred Dorset rams. In that case, on average, 50 percent of the lambs would be expected to be B+. Then Janet retains twice as many ewe lambs as she needs for replacements, and they are bred. Next lambing, those who produce twins are assumed to be B+ ewes and those who produce single lambs are assumed to be ++ ewes and culled. The procedure is not 100 percent accurate, but on average 90 percent of the retained ewes will really be B+ ewes. DNA testing can be used to produce precise analysis of the ewe's B-gene status, but the cost is not justified in most instances.

In Janet's own words: "The nifty thing about this gene is it can be introduced into any breed or type of sheep, and after back crossing with your fa-

Here is a Tamarack Farm mostly Dorset ewe (who has one copy of the Booroola gene) with her triplets. Email tamarack@pinenet.com for information.

vorite breed for a number of generations, you will now have a sheep that exhibits the traits of your favorite breed, but now produces nearly one more lamb than it did before."

Out-of-season breeding honors go to the Dorset and its crosses, with Rambouillets and Merinos coming in second. Some of the newer breeds such as Morlam and Polypay are also highly touted in this arena. Management schemes that include year-round breeding make better use of facilities and labor and even out the work load on a sheep farm. Year-round lambing is especially well suited to flocks kept in confinement. You should be choosy about where you get Dorsets: there has been some introduction of genetics from other white-faced breeds into the "purebred" Dorsets, which diminishes their ability to come into heat in off-season times.

For the person who is completely new to sheep and can afford to experiment a bit, I would recommend buying a few reasonably priced crossbred ewes of known ancestry to learn from. It is foolish for the novice to invest a lot of money in registered stock. In any case, crossbred ewes generally perform better on all counts than purebreds; this has been demonstrated by the long experience of producers and has been confirmed scientifically by J. A. Vesely's research at Agriculture Canada's Manyberries Research Substation in Alberta. Crossbred vigor (or heterosis) is a fact, and crossbred ewes will cost less.

As to age of the ewes you purchase, I suggest that you get ewes who have already lambed at least once but are not over four or five years old. Lambs are pretty wild and skittish and tough for the beginner to handle, especially at lambing. At the other extreme, an old ewe with half her teeth gone needs the experience in handling that the new shepherd does not have. A healthy two- or three-year-old is an ideal beginner's ewe.

Buying Ewes

Many people buy their first sheep at a local auction, have big problems with them, and are soured on sheep forever. Most stock at local auctions are the culls from flocks and should go into the freezer rather than into a breeding flock. You are much better off buying from an ethical breeder. Expect to pay about 150 to 200 percent of what a 105-pound market lamb is worth for a quality crossbred ewe of one to three years old. You may find bargains if you nose around a bit. For example, we cull good ewes from our flock every year if they don't produce twins. They are otherwise excellent sheep, and we sell them at a bargain to producers who understand why we are getting rid of them.

If you don't want to deal with a ram at first, bred ewes are available in the fall in most parts of the country. Usually the seller knows the parentage, at least to some degree, and some will even guarantee that the ewes are bred. If you know and trust the seller, this is a good way to start because supporting a ram for a small flock is poor economy. Be sure the seller really knows the time of breeding so you can plan for lambing.

When buying sheep, your best bet is to take an experienced person with you. Offer to pay for the help, as you would for any business consulting. While you are at it, consider that person's advice on what kind of sheep to start with.

Lacking an experienced person's advice, the novice can check quite a few factors with confidence. The sheep should look alert and healthy. Avoid ewes with runny noses, inflamed eyes, untrimmed hooves, scruffy wool, swayed backs, or other obvious defects. Check the mouths. A sheep should have all eight incisors (the front teeth) on its lower jaw. These incisors should meet the pad on the upper jaw squarely, not touching behind it or hanging out in front. The molars should not be worn down to the gums or decayed. The age of a young sheep can be judged from the incisors. A lamb has a full set of small teeth. At a year, the center two are replaced by larger teeth. Then each year another pair of the permanent teeth is added, one on each side of the previous ones, until at age four the ewe has a full set of large permanent incisors. After that time the gums recede, and teeth are lost and broken as the sheep ages.

You might want to have a veterinarian look over the prospective flock members, although the cost might be more than you could justify. If you do use a veterinarian's services, consider having the vet draw blood samples (5 cc) from each ewe, and have them tested for a disease called Ovine Progressive Pneumonia (OPP), unless the source flock is certified OPP free. The preferred test is called a PCR test and is done at Colorado State University (questions should be directed to Jane Carman at 970-491-1281 or jcarman@vth.colostate. edu) as well as some other locations. This disease has no symptoms in young ewes but will seriously affect the flock as the sheep get older. This is a useful precaution in order to avoid starting off a new flock with one or more carriers; there is no known treatment for the disease.

Another chronic disease in sheep is scrapie. It is a slow-developing, wasting disease that is difficult to diagnose. The USDA's Animal and Plant Health Inspection Service (APHIS) is attempting to eliminate scrapie in the United States. This will involve inspection and certification of flocks as scrapie free, with the hope of eventually eliminating the disease. The program is in its infancy but will become more important. A link to keep informed is at www.aphis.usda.gov/oa/pubs/fsscrapie.html. Also check with your state veterinarian to see if a program exists in your state. Eventually, having a certified scrapie-free flock will increase the market value of breeding stock from your flock. There is a genetic test for determining how susceptible a sheep is to scrapie. Check with your veterinarian. Tests for the presence of the disease are in the works too. All else being equal, if you are able to buy sheep from a certified scrapie-free flock, go for it.

Check the udder to be sure it has two teats. Also, feel the bag with your hand and fingertips to see if there are lumps or hard regions. The whole bag should feel soft and pliable. The ewe's body should be free of lumps on the skin. Be sure she isn't blind. Check the feet for any evidence of foot rot or other lameness; any soft or odorous parts should be viewed with suspicion. Listen to the breathing by putting your ear to the sheep's nose to see that the lungs sound quiet, not rattly. When you buy the sheep, get a guarantee of sound health if you can. (Some sellers of bred ewes will offer to replace any that do not lamb, and others may offer to replace a ram that proves to be infertile. However, as a general rule, it is "buyer beware.")

The most reliable source of breeding ewes is a sheep raiser who has been around for a long time and has built a reputation for honesty and quality stock. That person may be hard to find, but do some asking. Breeders who advertise consistently in trade publications will be careful to protect their reputations and are more likely to be trustworthy. One good place to buy sheep is at a regional sale held by an association or other group. The sale managers usually have a screening program for both the sellers and their sheep in order to protect the reputation of the sale; this gives the buyer an extra measure of protection too.

Buying a Ram

When buying a ram check the same areas you would with a ewe, except that in lieu of feeling the udder, feel the testicles. They should feel firm but not hard, and there should be no prominent lumps at the bottom. The testicles should be large, and both of them should have descended into the scrotum.

If you are buying a ram lamb, take along a tape measure. A study reported in *Iowa Veterinary News* showed that at an age of 150 to 160 days, infertile rams had a scrotal circumference of 25 cm (10 in.) or less. Fertile rams had scrotums with a circumference of 32 cm (12.5 in.) or more at the widest part. Scrotal size is no guarantee of fertility, but it is an important indication.

The ram should have a masculine appearance, at least to some extent. A ram with prominent, muscular shoulders will look very masculine but may also produce offspring that are built like him, and the packers want lambs with small shoulders and big rear legs. Get a ram with large rear legs and a wide loin. Put your outspread hand across the loin. With a yearling or older ram it should be as wide as the distance between your thumb and middle fingertip (about seven to eight inches).

If you can afford it, get a ram from a line that has the qualities you want. That means buying from a breeder who keeps good records and has a reputation for producing quality rams. For most shepherds it does *not* mean buying a fancy registered animal from a purebreeder who is big on the show circuit. Excellent rams can be had for a far more acceptable price from commercial sheep breeders who raise their own replacement rams than from the breeder with the $20,000 ram named Mr. Wonderful. If you do buy a purebred ram, ask if the flock is enrolled in the National Sheep Improvement Program (NSIP). See www.nsip.org for up-to-date information. Don't overlook cross-bred rams in your searching. The hybrid vigor effect is not quite as strong as the so-called first cross, but a crossbred ram on a crossbred ewe has a slight advantage over the purebred ram, all else being equal.

Choose rams carefully, because, unless you are shipping all lambs to market, the genetics of the ram will persist in your flock for a long time in the replacement ewes you retain. Some producers use one ram for market lambs and a different one for replacement ewes.

Try to get a guarantee of fertility from the seller, but don't really expect it. You can buy a yearling that was used as a lamb, although that doesn't really assure that he will still be fertile as a yearling. We always use retained ram lambs on a handful of ewes their first year so that we or a prospective buyer have some idea of their potential. Also, one doesn't have any way of knowing how many ewes a ram can service until he has been tried, although large testicles are strongly suggestive of a high sperm count and sperm supply. Many breeders use about one ram for every 20 to 25 ewes. One year we used a Finn × Lincoln ram on 90 ewes and he settled every one; the ewes produced more than a 200 percent lamb crop, which suggested that his sperm count was high all the way through.

All new stock should be quarantined for at least a month if possible. Be alert for any signs of foot rot or other disease. It would not be out of order to vaccinate for vibriosis, sore mouth, or other diseases of local importance. Don't introduce a carrier into your flock if you can help it. Treat all new stock for internal and external parasites (keds, lice) before putting them with the main flock.

Nutrition

For sheep raisers who operate on a conventional schedule of once-a-year lambing in late winter or early spring, the rebuilding time is an interval when pastures are used as the principal or sole source of feed for the ewes and rams.

Nothing like a midday snooze with a few friends to make a hot summer day tolerable.

Actively growing grass and other pasture plants are a highly palatable and nutritious feed for sheep, and they'll begin to restore their depleted bodies in short order. In fact, the shepherd should be alert to the possibility that some of the ewes will become overweight during the rebuilding time. Once they are at optimum weight, they require the equivalent of only about two and a half pounds of hay per day to maintain that weight. The shepherd's job is first to make sure that gains are being made and then to guard against some of the flock getting too fat.

Pasture Feeding

There are a number of ways to adjust the amount of feed to a flock on pasture. The easiest way is to have the size of the flock and the size and productivity of the pasture matched in such a way that there is a stable equilibrium between plant growth and sheep grazing. This approach is favored by many, and some studies support the idea that pasture utilization is most efficient when the sheep are just left alone to eat at will. In less than ideal situations the flock may become overfat and underutilize the pasture; or, conversely, they can be left short of feed and may destructively overgraze.

Another way to use pastures is to control the amount of time that the sheep are allowed access for feeding. Some estimates indicate that the carrying capacity of a pasture can be doubled by this method. This gain in efficiency of utilization is made at the cost of the labor required to move the flock to and from their active pasture and at the expense of providing a drylot area for the sheep when they are not feeding. In addition, a large share of the manure and urine accumulates in the yard, where it is a nuisance, instead of on the pastures, where it is needed. The disadvantages notwithstanding, I like the limited-access approach. The shepherd has a daily opportunity to observe the sheep, and they remain accustomed to being handled and moved about. They also appear to utilize the varied flora of the pasture in a much less wasteful way. Every shepherd will notice that sheep prefer to feed on the part of the pasture where they have been feeding previously, ignoring other parts. They do so because they find actively growing plants more palatable than mature ones. The sheep will ignore the mature grasses and herbs and clip a few areas until they graze them almost out of existence. Overgrazing is avoided when they are given only a few hours a day in the pasture. They learn very quickly that they have a very short time to fill their rumens, and they stop being picky. They eat everything in sight as fast as they can put it away. As a result, the pasture gets grazed down almost everywhere more or less equally. The sheep fill up in a short time and are readily moved to a lounging area.

The move to a drylot is made especially easy if no water or salt is provided in the pasture. The sheep will know that plenty of fresh water is waiting for them and will need little urging to leave the pasture, especially in the heat of a summer day. Provide some shade for them, and they'll be content in their yard until grazing time the next day.

Many producers have found that they can not only double their pasture's carrying capacity but even take a cutting of hay from a pasture before turning the sheep in. Removing a hay crop not only makes good economic sense but also, if done early in the season, gets rid of old, mature plants that the sheep would ignore if given the choice.

Sheep and cattle can also be grazed together. That practice often increases utilization because sheep and cattle eat different plants. Studies have shown increases of 10 to 30 percent in pasture output with mixed species grazing.

If the flock is subdivided into groups that require different amounts of food, the limited-access method permits the shepherd to put out each group at different times for different lengths of time. Alternately, supplemental hay can be provided in a drylot for the sheep that require it.

If the rebuilding time does not coincide with pasture season, the ewes are easily maintained on hay or silage. About two and one-half pounds of average quality hay (10 percent protein) will sustain an average-sized (150 lb.) ewe. If rapid rebreeding is to be attempted, feed can be increased to give a steady weight gain until the breeding condition is reached. In this case, the feeding of grain or a pelleted supplement might be considered. A ewe can eat only at most 3 to 3.5 percent of her body weight in hay each day, so potential gains are limited by her filling up. A concentrated feed allows her to get more nutrients into her body and to recover from lactation more quickly.

Grain Feeding

In summer, the sheep raiser should take note of grain prices and crop prospects. Prices fluctuate over a fairly wide range, and the shepherd should try to buy cheaply. Prices are reported in newspapers or on radio or TV in rural areas. You'll find that the price you have to pay a neighbor or the local elevator or feed store will not be the one listed in the reports because of a cost differential reflecting the expense of transportation or lack thereof. In grain-producing regions the prices will be generally lower than the quotes from terminals.

Consideration should also be given to income tax angles because it might be advantageous to buy grain before the end of a tax year in order to balance income against the expense of the grain. You need not have on-farm storage to do that in many areas. You can buy grain and store it at the local grain elevator for a fee. If the elevator has excess storage capacity, it will probably cost you less to store it there than to build your own facilities. Once again, everyone has to do some figuring, using local prices. You don't actually have to take any grain to the elevator for storage. You simply buy grain from the elevator when you think the price is right and pick it up as needed. Needless to say, any grain stored at an elevator is mixed with other grain of the same type. You won't get your own grain back when you pick up a truckload a few months later, so if your grain is a special kind or is grown in an uncommon way— without the use of chemicals, for example—you'll have to store it elsewhere to be sure of getting the same grain back. This is especially important for sheep raisers who are trying to sell organically produced lambs, because they must be able to prove that the lambs' feed meets organic criteria set by a state regulatory agency or by a specialty buyer.

Feed Requirements

Probably the most common error made by sheep raisers is to use the wrong amount of feed or the wrong kind. Obviously, feeding too much is costly because of waste, and feeding too little results in poor performance of the sheep and lambs. In addition to quantity of feed, the ration must be balanced with respect to the energy content (expressed in millions of calories), the amount of protein, and the amounts and proportions of vitamins and minerals. The requirements are unequal for different stages in the ewe's cycle and are not the same for growing lambs or for rams. Calculating a suitable ration is not difficult. The National Research Council's book *Nutrient Requirements of Sheep* contains tables that list the feed requirements for sheep of a variety of weights and at different stages of growth. One section describes overall nutritional needs of sheep, and tables of composition of common feedstuffs are given. Using the tables, one can calculate what combination of available feeds meets the tabulated needs of the sheep. A portion of the NRC tables is reproduced in appendix 5, and examples are given to illustrate the arithmetic involved in the calculations. The complete book can be ordered or even read at www.nap.edu/books/0309035961/html/index.html.

I will include feed recommendations for each stage of the ewe's cycle. These are based on average-quality hay and shelled corn to give some rough guidelines, but each shepherd should spend some time with a pencil, paper, and calculator to evaluate his own sheep-feeding program. An important principle to keep in mind is that the growth and general health of the animal will be governed by that part of the total diet that is in shortest supply. In other words, if the diet contains adequate or even excess protein, vitamins, calcium, and phosphorus but is deficient in carbohydrates, then the sheep will not thrive, performance being limited by a lack of calories in the diet. They need what the feed salesman calls a balanced diet.

If you are tempted to buy complete feeds and supplements from a feed salesman, take the time to sit down and calculate what it would cost you to provide the same levels of nutrition from a homemade mixture. I think you will decide to prepare your own feed unless the convenience factor is of overriding importance. There are no magic ingredients in the commercial feeds. If you have no grinding or mixing facilities, you can have your own formula put together at a grain elevator or feed store for a nominal price.

Salt

In summer you do not need to provide your sheep with anything but pasture for adequate nutrition, unless the ewes are in late gestation or lactating. Sheep like salt, which should be given free choice in loose form because sheep can break their teeth on salt blocks of the common type. The salt can be either iodized or of the type called trace mineral (TM) salt. Many shepherds mix the salt with an equal amount of dicalcium phosphate to provide supplemental calcium and phosphorus.

It is very important that sheep have access to salt at all times. Have an adequate number of salt stations and keep them filled. Salt is an essential part of a sheep's diet. Don't try to save money by withholding salt. The commercial weather-proof feeders are probably the best investment. Sheep will consume ¼ to ½ ounce of salt mix per day per head. The mixture does not need to contain molasses or protein.

Most TM salts also contain copper because they are designed for cattle rather than for sheep. Unfortunately, sheep have a limited tolerance for copper and can be poisoned by an excess. There are special copper-free mineral mixes made especially for sheep, but they can be costly. You are possibly safe in giving some ordinary TM salt to sheep, free choice. I have used one part TM salt to three parts white salt with no problems. Never mix a copper-containing salt with feed. To include salt in feed, use plain salt or iodized salt.

Trace Elements

If you are concerned about the copper levels in your sheep and their feed, you can have some analyses made. Your veterinarian can give you the address of a diagnostic laboratory that can do the analyses. You will need samples of the liver from a few sheep and some representative specimens of feed from your farm, including pasture grasses. You need not kill a number of sheep just to get liver samples. Just save half a dozen fist-sized portions of the livers from sheep that either died or were butchered.

The copper level in the livers should not exceed about 100 to 500 parts per million (ppm). If your results are above that level, you should try to discover the source of the copper and eliminate it. The feeds should be expected to contain about 5 to 10 ppm copper. You might even find that your sheep are deficient in copper, a rather uncommon situation in North America but one that can be brought about by an excess of the trace element molybdenum in some areas, notably in western Canada. Molybdenum interferes with the absorption of copper by the sheep. Conversely, extremely low levels of molybdenum can generate excess copper absorption.

If you experience copper toxicity, instead of treatment, focus on trying to find the reason and source and remove it. Surviving animals can be treated with ammonium molybdate in their feed for three weeks or so. Valuable animals can be treated with an expensive human drug called D-penicillamine to help get copper out of the liver. Consult your veterinarian.

Another trace element that the shepherd should be concerned with is selenium. In many areas of the western plains and Rocky Mountains selenium is sufficiently abundant in soils as to cause toxic levels in forage plants. The toxicity of selenium has been known since the days of alchemy, but it was only in the 1950s that some veterinarians in the Pacific Northwest discovered that selenium was an essential nutritional element, and that a deficiency in selenium in the diet had highly adverse effects on sheep and other species. Since then, shepherds and veterinarians have found that selenium deficiencies exist in many parts of the United States and Canada. These deficiencies are readily corrected by providing supplemental selenium in the form of sodium selenite or selenate mixed with free-choice salt or as part of a feed mixture. The legal maximum level of selenium in a salt mixture is 90 ppm (or grams per ton) selenium.

How do you know whether you have a selenium problem? There is no simple answer, because the distribution of selenium in rocks and soils is so spotty and localized that one cannot define even broad areas where excesses or deficiencies can be expected for certain. While one suspects that selenium deficiencies exist in many places—including the Pacific Northwest, northern California, the northern Midwest, New England, and Florida, as well as much of Canada—there is no way to know in advance whether or not a problem exists on a given farm, and neighbors may face quite disparate situations. One solution is to ignore the problem until deficiency symptoms appear. As I will mention again in the chapter on lambing, selenium deficiencies result in weak lambs and a condition called white-muscle disease in which parts of the muscle tissue die, crippling the lamb, and ultimately resulting in its death. This disease is an indication of a selenium deficiency, commonly found together with a lack of vitamin E.

For the curious who want to spend the money, blood samples can be analyzed by a laboratory to check on selenium. The blood should contain about 0.1 ppm selenium, with levels up to five times that amount considered quite safe insofar as toxicity is concerned. Levels below 0.02 ppm will result in clear clinical symptoms such as crippling, and less obvious effects can occur for any levels below 0.1 ppm.

Most other trace elements seem to be present in satisfactory amounts in pastures and in hay from nonirrigated land. Feeding supplements of manganese, iron, cobalt, zinc, and other elements is probably not critical, although each element appears to have one or more important roles. Iron is abundant in most feeds, though I'll mention the need for a supplementary source for lambs later. Cobalt deficiencies are common in parts of England and Australia but are unusual in North America. Sheep mineral mixes provide these elements, as does TM salt.

In inland parts of the country some iodine supplementation is appropriate. Sheep mineral mixes, TM salt, or iodized salt can be used. Be sure to use ordinary iodized salt with 0.01 percent potassium iodide (which translates into about 0.005 percent iodine) and not one of the medicated iodine salts that contain much more. An iodine content of up to 0.02 percent is recommended by some and is perfectly safe.

It is important to understand that there are no hard rules as to what trace elements (minerals) are needed by a given flock in a given region. Different breeds need different amounts, and the soil chemistry and feed compositions are different in different regions.

ENVIRONMENT

Fencing

I'm a firm believer in keeping the sheep outdoors as much as possible. During a sheep's rebuilding period, pastures are the usual feed source, and except for the western ranges, that means fences. The old saw about good fences making good neighbors makes a lot of sense. Your sheep grazing in someone else's crops or garden may create a lot of havoc and can cost friendship or money. Not only that, if your sheep eat a neighbor's alfalfa, you'll not only strain a friendship, but you may lose some sheep to bloat as well. Also, a fence that lets sheep out will just as readily let dogs in. Dogs can maim and kill a lot of sheep very quickly, and losses can be disastrous. The solution is good perimeter fencing. A sharp eye and a rifle might be a solution, but local laws and customs, as well as relations with neighbors, must be considered.

Besides, a shepherd cannot be watching the flock all of the time. Fences are better.

There are as many styles of fences as there are brands of candy bars, but some principles apply to all of them. They must be close enough to the ground to prevent animals scooting under, in either direction; they must be high enough to discourage sheep, dogs, or coyotes from jumping over them; and they must have wires, boards, or other fencing material spaced close enough to keep animals in or out.

Wooden fences are built by shepherds at both ends of the economic scale. The wealthy build neatly painted board fences or picture-book split-cedar rail ones. The poor build fences of cut poles, brush, old stumps, and anything else that can be scrounged for free. Most sheep, however, are confined by wire fences of one sort or another. Fences made only of barbed wire are seldom effective for sheep because they can usually push between the wires and make good their escape. The shepherd's choice is really between an electrified fence and one that combines wire mesh and barbed wire.

According to the people who sell them, electric fences are cheaper than other types if one compares the cost of all-new materials. Electric fences typically have one part of the electric circuit as the ground or earth and the other, or "hot" part, as the fence wires, so a sheep or predator touching the wires will feel a shock from the electricity passing through the body. The electricity is provided by an energizer that delivers brief pulses of thousands of volts to the hot wires. The energizer is powered by power line electricity or batteries, and some even have solar panels to keep batteries charged. Line sources are generally preferred if power is available. The ground connection to the charger should be very good, usually made of several long ground rods that reach moist soil or earth at depth. The preferred type of energizer is called a low-impedance type. Older fence "chargers" are not suitable. Popular brands are Intellishock, Speedrite, Gallagher, and Maxishock. Be sure to include lightning protection devices in your design, because a quick lightning strike can destroy an expensive energizer in microseconds. Sellers of electric fencing equipment usually have excellent booklets about how to build good fences. See Premier Sheep Supplies, p. 309.

Fence posts must be of non-conducting material, or insulators must be used to hold the wires. On the other hand, savings are made because smooth wire is cheaper than barbed, no wire mesh is needed, and the fence posts can be placed much farther apart because the type of wire that is used (called high-tensile wire) is very strong and can be stretched very tightly so as to prevent sagging. The advantage of wide post spacing is largely lost on uneven ground, and the choice of electric or nonelectric fencing must be made with the terrain of the farm in mind, because advertised cost figures assume smooth ground.

If the soil below the fence becomes very dry, it will no longer conduct electricity, and the fence loses its effectiveness. A solution to this problem is to make the fence of alternating ground and "hot" wires, so that no matter where an animal pokes its head through, it will receive a shock. There should be a ground wire down at almost ground level to shock animals trying to go under the fence during dry weather.

Electric fences have a couple of advantages other than cost. They seem to discourage dogs and coyotes more than a plain wire fence, so if predators are a problem in your area, this is a big plus. Another advantage is the almost unbelievable fact that one doesn't need a gate in order to drive a vehicle through the fence. The long, unsupported sections of the fence have stays that keep the wire spacing correct. If the front of a vehicle is smooth, it will push and rotate the stay and wires flat to the ground as the vehicle passes over, after which the fence springs back into place. I recently read of a driver in the West who woke up after losing control of his car only to find himself in a fenced pasture. Seek as he did, he couldn't find a gate, and both he and the law enforcement officials decided that he must have sailed over the fence and miraculously landed unscathed in the field. Closer inspection showed that he had just driven through a high-tensile wire fence that snapped up behind him.

Disadvantages of electric fencing include the statutory requirement in some areas that it be marked with warning signs, and the hazard that older types of fence chargers may start fires in dry grass or brush that touches the wire. That is not a serious problem with low impedance energizers. Fences can lose their shocking ability if the wires become shorted by contact with conducting materials, including wet grass, but attention to trimming the grass can prevent that problem. Also, low impedance chargers are not as strongly affected by such "shorting." The power source can be interrupted or a battery can go dead, so routine checking is useful. One should use monitoring lights and a special voltmeter for checking. In an emergency just hold a three-to-four-inch piece of grass in the hand and use the grass to probe the fence. You'll feel a mild shock through the grass to tell you the fence is functioning.

Be aware, too, that high-tension fencing is very unforgiving of sloppy workmanship. The pull of the taut wires on fence corners and intermediate posts and braces is very great, and the supports must be placed deep in the earth and be well braced. Use the suggestions for electrified sheep fences available from dealers in fencing supplies.

Temporary electric sheep fencing is also available. It is made of plastic mesh with embedded wires and is attached to lightweight posts that push in easily. This fencing can be erected and taken down quickly, and, while it is relatively costly, it offers a convenience and flexibility that justify the expenditure. Temporary fences can be made around grazing areas that otherwise would be unusable, small groups of sheep can be separated, and other uses may come to mind once you think about the possibilities. I used this sort of fencing for years with no mishaps. Others have had sheep tangled in the fence, and all sorts of disasters. I guess it depends on the sheep. Under no circumstances depend on such fencing to keep a ram from the ewes.

Wire Mesh Fences

Until the advent of modern electric fences, wire mesh was the fencing of choice for sheep. The wire mesh is usually the type called woven wire, available in heights of 26, 32, 39, and 47 inches in most brands. A typical fence consists of a single strand of barbed wire an inch or two off the ground, followed by woven wire, usually the 32-inch variety. This is followed in turn by another strand of barbed wire an inch or two above the woven wire and is topped off with additional strands of barbed wire at three-to-five-inch intervals to the top of the post. A height of four feet for the top wire will discourage, though not prevent, predators from jumping in.

Woven wire is available in two spacings of the vertical wires, six and twelve inches. The twelve-inch spacing will not confine lambs or keep out small predators. However, it is the choice of many shepherds because the sheep can exercise their usual propensity to eat through the fence and still be able to withdraw their heads without either strangling or being stuck for a long time. The smaller stay spacing will confine lambs, but now and then a sheep will get caught or killed because, although it can jam its head through, it is often incapable of figuring out how to get it out again. We have lost only one ewe to strangulation in a fence, but I have pulled plenty of pretty tired ones out. Lambs get stuck frequently as they grow up, but most of them eventually learn to stay out of fences or finally figure out how to extract themselves without assistance.

My own preference is to use the six-inch stays for perimeter fencing and the twelve-inch for cross fences. That way there is maximum security between the sheep and the outside world, and the risk of a sheep's strangling is balanced by the increased safety of the flock from predation or escape. The twelve-inch mesh for cross fences lets the lambs roam a bit, but they seldom stray far from their mother, so that usually is not a problem. For interior fences I prefer the 32-inch woven wire with no barbed wire at all. If the sheep are well fed, they will make no attempt to jump the fence, and it is very convenient to be able to step over the fence without tearing pants on a barb or two.

The cost advantage that is claimed by the electric-fence advocates is often reduced by the fact that one can obtain used fencing at auction sales, both steel posts and wire being common items. The buyer should be wary of rolled-up woven wire, though, as the core may be junk. Posts are easier to check, and bargains can be had. I acquired hundreds of used steel posts at half the new price by placing a single classified ad in an area advertiser. If the posts are rusty the rust will migrate to your new fence wire and shorten its life. Paint the posts with a good rust-preventive paint where the wires contact the post. The rust problem is an argument in favor of wooden posts with heavily galvanized staples.

Corners and intermediate bracing should be sturdy, though they need not be as rugged as those for high-tensile fences. Tightly stretched fences will remain useful for years if made properly. Studying good-looking fences or working with an old hand will help you learn the tricks of the trade. Even the tightest fence needs periodic maintenance, especially in snow country where the settling of the snow pack through the winter leaves the wire dragged down in festoons when the snow melts. A fence tightener that restores the crimp to the wires will tighten a sagging fence quickly and easily in the spring.

Gates

Gates for either sort of fence are limited only by the builder's imagination and pocketbook. Purchased aluminum gates usually have openings that a small sheep can squeeze through, but woven wire can be attached to them or additional metal strips can be added. Combinations of woven wire and boards make attractive and durable gates. For seldom-used openings the simple Western gate or Texas gate made of woven wire or woven plus barbed wire with skinny wooden posts at each end, held in place by wire loops, works well, although children or lightly muscled persons may have trouble opening and closing them. Welded steel cattle panels or hog panels make effective gates too. For electric fences the gate is usually not electrified, the hot wire being an insulated cable that is buried underground across the gate opening.

Pasture Ecosystems

The pastures that are enclosed by the fences should be treated as a valuable asset and taken care of almost as well as the sheep that use them. As mentioned earlier, pastures can be utilized in continuous grazing, or sheep can be moved, or rotated, in and out of individual pastures. There are arguments in favor of each method, but some general principles apply to most situations. An overriding consideration is the strongly seasonal growth of most grasses. Cool-season grasses—bluegrass, quackgrass, bromegrass, and crested wheatgrass—have their main growth period in spring and early summer. After the rapid early growth they go to seed, and for practical purposes almost stop growing for the balance of the season except for a short growth period in the cool days of fall. In some parts of the country there are warm-season grasses such as switchgrass, indiangrass, bluestems, sideoats grama, and sudangrass, for instance, that have their maximum growth in the hot part of the summer. A few other species such as orchardgrass and reed canarygrass grow more or less throughout the season. The legumes, such as alfalfa, clover, sweet clover, and birdsfoot trefoil, grow all season as long as sufficient moisture is available.

The net result of the seasonal growth pattern of most grasses is that no pasture that is populated by a single species or by a number of species of similar seasonal growth pattern can be expected to provide summer-long grazing. It will be underutilized during the peak growth season and overgrazed at other times of the year. Continued occupation of such pastures at high stocking rates will kill off the grasses or at least thin them to the point at which undesirable species will fill the gaps and compete for water and nutrients with the valued species. This is particularly true for the warm-season grasses and some of the legumes, especially alfalfa.

Heavy grazing without an interval for regrowth will kill off forage plants quickly. The native warm-season grasses of the Great Plains, after having provided buffalo feed for millennia, were largely wiped out by heavy stocking of cattle by cattlemen. Cool-season grasses withstand heavy grazing better, but even they suffer.

It is important to keep in mind that the quantity of roots of a grass is proportional to the amount of plant material above ground. With continued close grazing the reserve of food in the roots is depleted, and with tops constantly taken off there is no resupply. The root mass shrinks in both size and number of rootlets. When grazing is stopped, regrowth is slow because the plants must rebuild their mass by manufacturing food from scratch rather than calling on reserves held in the roots. If heavy grazing is continued right up to winter in cold regions, there will be a heavy winter kill of the roots.

The only way that continuous grazing can be nondestructive is with low stocking rates. Either a small number of sheep are placed in a pasture, or the sheep are moved from time to time to lower the effective stocking rate. Pasture rotation at its simplest uses two pastures, leaving the sheep in one while the other recovers, rotating as often as needed. This system can be improved upon by using three or four pastures, rotating in sequence among the various units, all of which have more or less the same species of forage plants. Rotation helps not only by giving plants a regrowth period; it also cuts down on the population of internal parasites such as worms because many of the larvae in the grass die off before being ingested by a sheep host. The larvae are at a special disadvantage in an eaten-down pasture because they are exposed to sunlight and drying winds without the protection of plant cover.

Grazing should also be timed according to the growth habits of the grasses. Jointed grasses like bromegrass, quackgrass, and some of the wheatgrasses have a growing point that is near the surface of the ground in the early stages and that moves upward as the grass grows. If the growing point is removed too early by grazing or mowing, the growth is effectively stopped. On the other hand, if the grass is allowed to go to the so-called boot stage, when the seed head has formed inside the stem but has not emerged, then cutting or grazing encourages rapid regrowth. If the seed head is allowed to emerge, dormancy is induced.

Bluegrass, junegrass, orchardgrass, and Russian wildrye form only a few jointed stems that develop seed heads. The growing points of the other stems stay near ground level, and heavy grazing doesn't stop growth unless root reserves are seriously depleted. In other words, the latter group of grasses produces the most if grazed more or less continuously during their active growing phase. The other group should be allowed to go to boot stage before removal of the tops.

A more elaborate system of rotation can be devised if the species composition is not uniform. For example, if one had a bluegrass pasture and a bromegrass pasture, good timing of their use would maximize their yield. The bluegrass pasture could be grazed first and grazed continuously until the bromegrass reached boot stage. Then sheep could be turned into the bromegrass to eat it down to a two- or three-inch length. At that point, the sheep could be put back on the bluegrass until the bromegrass reached boot stage again, and so forth, as long as the cool weather lasted.

For midsummer, the warm-season grasses can be planted in some areas, but seed is expensive and establishment difficult. Also, these grasses will be retarded or killed by early grazing. An alternative is to plant orchardgrass or a legume. Orchardgrass-alfalfa mixtures are popular, or birdsfoot trefoil alone can be used. Pastures of these species can be grazed during the hot months, provided they are not grazed continuously. A period of regrowth must be allowed after removal of the tops.

On our farm we use a bluegrass to quackgrass-bromegrass rotation for early grazing. Then the flock is alternated between an alfalfa-orchardgrass pasture and a birdsfoot trefoil pasture for the warm season, with a return to the cool-season grasses for a short period in the fall. I usually can get a cutting of hay from the trefoil and orchardgrass-alfalfa before the sheep get to them. With this plan one can graze ten ewes to the acre and also put up some hay from the same pastures, usually about a ton to the acre in years with enough spring rain. This sort of approach is applicable to the northern tier of states and Canada with minor modification. In the southern parts of the country grazing seasons are extended, and different plant species are appropriate. Suggestions from soil conservation service and extension personnel are very helpful for designing a grazing program using suitable regional species.

Pasture Modification

Existing pastures can be improved by a variety of methods. The most obvious is reseeding. If a pasture can be tilled to prepare a seedbed, then virtually anything can be planted by conventional methods. Clean tillage is a necessity with some species such as birdsfoot trefoil, which competes very poorly with other plants. Once established it does very well and will reseed itself if left to go to seed once a year, preferably in the summer rather than fall. The sheep will plant the seed with their feet when they graze the pasture later in the year. Properly managed, trefoil will take over a field when once established.

If tillage to a clean seedbed is not possible, a pasture can be intentionally overgrazed, then seeded after breaking up the surface with a disk or toothed harrow. Seeding rates should be roughly doubled because much of the seed will not germinate. Small seeded plants such as alfalfa and clover can be seeded in this way. Grasses with large seeds, such as orchardgrass, are more difficult to get into the soil without a seed drill. There are also seed drills that are designed for planting seed into existing sod, and in some areas these may be available on a lease basis from a soil conservation district or some similar organization. Many of these drills apply a herbicide at planting to reduce competition for the new seedlings. Herbicides can be applied or burning used to reduce competition before planting.

Pastures can be changed in character—without planting anything—by the timely use of grazing. If a given species is to be favored, the pasture should be grazed hard during or just prior to its period of maximum growth. For example, if a pasture was mixed bluegrass and quackgrass, and you wanted to favor the quackgrass, the sheep should be put in the pasture when the quackgrass is in boot stage and the pasture grazed heavily. When the sheep are removed, the pasture should also be mowed to get rid of any plants that the sheep missed. The quackgrass, or a similar species, will grow back rapidly and shade the other grasses and herbs, gaining an advantage for itself. If this is done year after year, the quackgrass will take over.

If one wanted to favor the bluegrass in the same pasture, grazing or mowing should be done before the quackgrass is in boot stage; this would send it into dormancy and favor the bluegrass, which would keep right on growing.

Warm-season grasses that are grazed only in midsummer will gradually take over from other species, provided overgrazing is scrupulously avoided. The warm-season grasses can also be encouraged by heavy grazing of pastures in the late fall when they have become dormant, but then the cool-season grasses continue growing to store up root reserves. This will leave the cool-season grasses short of food with a resulting winter kill and a slow start in the following spring for the survivors, giving the warm-season grasses less competition in the early growth stages.

Pastures can benefit from applications of fertilizer. Costs should be considered carefully before applying nitrogen because the expense is high. You will need to know typical responses of pastures to fertilizer in your area in order to make a rational decision. Either consult someone locally or do limited experiments of your own. Needless to say, cleanings from the barns and drylots belong on the pastures too. Fertilizer can also be used to aid in changing the species makeup of pastures. For example, quackgrass responds to an application of nitrogen much more than does bluegrass, so fertilization favors the quackgrass in the long run.

One of the best and least expensive pasture improvement and maintenance tools is a mower. If time permits in a busy summer schedule, a pasture should be mowed as soon as sheep are taken out of it. The mowing not only removes mature grasses that the sheep won't eat anyway, but also kills or retards weeds such as thistles, burdock, mullein, milkweeds, and many other nonforage species. I have managed to virtually eliminate mullein and Canada thistle from our pastures by mowing, but I'm still losing the war with another thistle species because I'm usually too busy to mow when I should.

Species with burrs that get into the wool, such as burdock, should be hunted down like escaped felons and uprooted and burned. The same is true for any poisonous species that may be present in your area. Sheep will usually avoid poisonous plants if they have sufficient other feed, but may eat them in hard times. I recall a friend in Kansas telling me of sheep raisers losing a lot of sheep to water hemlock in their area one summer when other feed was short because of drought conditions. Any undesirable plant can be eliminated by spot application of herbicide early in the season, but it is a continuous battle because of reseeding by airborne seeds, weeds in hay, and the like.

Summer Camp

Sheep can sometimes be sent away for the summer, as one would send youngsters away to camp. Some farmers have pasture for rent on a regular basis and will even water and care for the flock for a fee. All that is really needed is a fence to keep the sheep in and predators out, and a supply of water. Of course someone has to check on them once in a while, but summer is a time of few problems for the ewes. Every locality will offer different opportunities for putting the sheep out, and I know of sheep summering in such diverse places as the lawns of a municipal sewage plant and the outer reaches of military installations. Lawns around microwave relays, junkyards, government establishments, factories, corporate headquarters, schools, football stadiums, power substations, racetracks—any of these locations can be suitable for sheep. We put a small flock on a neighbor's place one summer at his request to keep some brush and weeds under control. Among the group was an older ewe with a chronic hacking cough who we figured would probably never last out the summer. She returned from her vacation as fit as if she had been to a health spa. The mixed diet must have contained something that was just what she sorely needed. That was years ago; she was a productive member of the flock for many years, and that's her face you see as the frontispiece of this book.

Rams are ideal candidates for sending off to camp. Find them a cool place with adequate feed and they'll be content. Rams are basically lazy, so let them loaf around away from the ewes, building up their sperm count to earn their keep come breeding season.

Dogs

Dogs are part of the sheep's environment for most shepherds. The dogs might be herding ones, guarding dogs, or predatory ones. In an ideal world, none of these types of dogs need be around, but in real life, one or more is part of the scene. Dogs and sheep are not necessarily inseparable in sheep raising. A friend of mine returned from a geological expedition to Morocco and observed that in that country the dogs stayed home and slept on the front stairs while the men trailed the sheep over the sparse grazing and the wives stayed home to do all of the work around the house. His theory was that the Moroccans could be pretty prosperous if the dogs were used with the sheep, and most of the men could stay home and do some useful work. I'm afraid my friend didn't consult with the Moroccan shepherds about this idea, nor does he understand the delights of sheep watching.

Herding dogs are almost as much a part of sheep raising as the sheep themselves. A good sheep dog is certainly the greatest labor-saving device I know of, with the possible exception of a corkscrew. One can raise sheep without a dog, but it sure is easier with one. The border collie is the classic sheep dog, and I personally would not even bother to try another breed. A border collie is born already programmed to herd sheep. All the handler has to do is teach a set of simple commands to direct what the dog wants to do instinctively. Teach the pup some simple commands, give it some love, and you've got a sheep dog.

Most border collies have a natural instinct to go away from the shepherd, encircle the sheep, and move them back toward the shepherd. With new pups, let them do that as you teach them some commands to go with their in-

A Sharplaninac puppy investigates a hay feeder in the pen it shares with some lambs. It will grow up to be a flock guard dog.

stincts. Be sure they learn commands to drop (such as Down!) and to stay there (such as Stay!), and enforce those commands. Then teach separate commands to go gather the sheep from your right and from your left. Most dogs will crowd too closely at first, but will usually learn on their own to give the sheep more room. After some weeks, one can start to teach the dog to herd along with the herder. That is contrary to their instincts, so be patient. One expert trainer comments that the average person trying to train the dog doesn't know how to make the stock move—and therefore is not qualified to tell the dog what to do.

Naturally, there will be individual dogs that are no good at all and others that are great. I'm also assuming that you get a dog from a line of working dogs, not show-ring types. Their whole interest is herding. They'll herd chickens, ducks, geese, soccer balls, small children, cats, anything. John Holmes in *The Farmer's Dog* recommends training them by using confined ducks, an idea I have not tried, but a good one. A sheep dog should be used frequently to keep it in good habits, but otherwise it is almost no trouble and is a pleasant companion as well.

When left alone, a sheep dog should be confined or on a leash, because some of them will herd the sheep on their own and can cause problems by pushing them into a fence corner or over a cliff, as described in Hardy's *Far from the Madding Crowd*.

Guard dogs are a fairly recent addition to the sheep industry in this country. They have been used for hundreds or thousands of years in southern Europe and adjacent parts of Asia Minor. The breeds have wonderful names, such as Great Pyrenees, Komondor, Anatolian, Akbash, Maremma, Kuvasz, Tchouvatch, and Sharplaninac. A guard dog's purpose is to stay with the sheep, warding off potential predators, usually coyotes. It is not a herding dog. Guard dogs protect sheep by patrolling, barking, scent-marking, and pursuing—even killing—a predator when the sheep are threatened. Guard dogs have proven to be very effective protection against predators in many environments, and are widely used even in range operations. In a survey in Colorado a total of 174 of 182 producers (96 percent) would recommend use of guard dogs to other producers. Any sheep raiser with predator problems should think seriously about getting a guard dog.

Donkeys and llamas are also used as protectors, but are really effective only in fenced pastures of less than a couple of hundred acres. With guard dogs or other animals, fencing costs can often be reduced, because predator exclusion is less of an issue.

The dogs are usually raised together with the sheep from about two months of age so that they form social attachments with them and work out their aggressive tendencies by protecting the flock. It is best to place a new pup with lambs, so it is not attacked by aggressive ewes. A guard dog should be taught to obey some limited commands, especially Down! Stay! and No! and learn to walk on a leash, but never treated as a pet. Feed the pup when you feed the lambs. Correct the pup if it chases or bites sheep, and praise it for desired behavior. If the dog leaves the sheep, return it to them immediately. Praise it for staying with the sheep. Guard dogs will be ready for guarding at six months to two years of age. If one guard dog is useful, you should consider adding more. In range conditions guard dogs work best with breeds of sheep having strong flocking tendencies, such as Rambouillets or Targhees, and not as well with less gregarious sorts such as Suffolks.

Most guard dogs will attack an unfamiliar human approaching the flock as readily as they would a coyote or a stray dog, so strangers must be warned to stay away, because these are big dogs that could do a lot of damage to a person. If you get a guard dog, be sure that neighbors know about it so some "helpful" oaf doesn't shoot it to protect your sheep, or a neighbor get attacked by your dog. Check www.ext.colostate.edu/pubs/livestk/publive.html and www.flockguard.org for current information.

Be sure that both herding and guarding dogs are protected against rabies by vaccination. In some states owners can buy and administer the vaccine, whereas in others a vet must do it. You can also consider getting yourself and anyone else who works with dogs and livestock vaccinated against rabies too. Even sheep can contract rabies.

Predatory dogs, whether wild or domestic, mean trouble for a shepherd. Coyotes, and even wolves, are a big problem in some parts of the country, but the domestic dog is the worst offender, especially in heavily populated areas. Our place is in a pretty much strictly farming area, but there are a few non-farmers and others who let dogs roam. A phone call has been sufficient to take care of dogs with known owners, although a load of number-six shot from a distance has a very salutary effect too. For dogs in packs, a loaded rifle is the only reasonable solution.

Check on local laws, and if they are not strong enough, work to get them changed. Dog owners must be made to understand that they are responsible for their dogs' actions. Proof is often difficult to establish unless the dog is observed chasing the sheep and then confined until the local law officials and the owner can be called. In our state, according to the laws, a dog in pursuit of farm animals can be shot. Owners almost never believe that their dog would harm anything, and a quick, accurate shot followed by a quiet burial is often the pragmatic solution. If you prefer not to kill, get a paint ball gun and mark the dog. The owner will have to admit that the orange-spotted dog was on your place.

If you have guard dogs, that will help, but be sure your neighbors understand that the guard dogs may kill their dogs if they stray into the property. This is a situation in which you need to be on good terms with the dog-owning neighbors. If you shoot their dogs, they may just shoot your very valuable guard dog in return payment.

You can carry insurance to cover losses from predators. If you insure your sheep, be sure that their value is established before you try to collect on the policy. Insure your guard dogs too. Most companies will try to pay as little as possible and may want to pay you only for what a cull ewe sells for at a local market. Itemize your sheep as individuals or by breed and assign a value to them that your insurance company accepts. If you get caught short with your insurance adjuster or are trying to collect damages from a dog owner, you can use your sales records to establish the value of your sheep. One raiser on the Pacific coast recently used asking prices for black sheep in *Black Sheep Newsletter* classified ads to establish the value of her sheep. A friend of mine in Massachusetts had moderately good results with the same technique. Good fences are still the best solution to most problems, although a neighborhood vigilante committee is sometimes needed. When someone in my area spots a dog pack, lots of phone calls are made to alert everyone, and rifles are loaded although we are the only sheep raisers for miles. Nobody likes a dog pack.

MEDICAL

Preventive medicine is the key to flock health, and the time between weaning and flushing is a good time to be sure that the flock is in fit condition. Good health of the sheep is encouraged by exercise, suitable diet, cleanliness, and prevention of infection.

In an ideal world the flock should be supervised at all times, but of course that rarely really happens in America, where shepherds generally do other things such as farming, taking care of other animals, working at an outside job, or writing poetry. The days when a shepherd and dog watched the flock on a round-the-clock basis are pretty much gone except in the West. Still, the shepherd should try to observe the flock at least once a day if possible. This check might consist of simply looking out the window at the sheep, or taking a break by walking over to them for a little relaxation. One reason that I favor moving the sheep to and from pasture each day is that it gives the shepherd two chances to observe them and their environment.

If you don't already know one, get to know a veterinarian soon after you acquire your sheep. Choose a vet who knows about sheep or is willing to learn. Some vets are interested only in dairy cows, or cats and dogs, so check around. Find someone whom you can work with and learn from, preferably someone you like. Keep in mind that you as a shepherd will have to learn to do a lot of veterinary care yourself, because the value of an individual sheep commonly does not justify the expense of a visit by your veterinarian. So, you'll need a vet who will teach you some techniques to do yourself, rather than insisting that only someone with a veterinary degree can do the job. Most vets are very good about this, but not all.

Work with your DVM to establish a schedule of preventive vaccinations and other measures to head off disease problems before they surface in your flock. Talk with your vet too about your nutritional program if you have any doubts. The veterinarian is a lot more objective than the representative of a feed company.

Throughout this book I will discuss some common sheep health problems, but don't get the idea that sheep are sick all of the time. They are fundamentally healthy animals that will rarely fall sick if treated sensibly. Always try to ward off sickness by prevention rather than waiting to fight a raging illness outbreak. Rebuilding is generally a time when very few health problems arise so it is a good time for you to learn.

Be sure to have a few basic health equipment items. First of all, buy a good thermometer. The electronic ones meant for people work fine with sheep and are inexpensive and accurate. You will also need a stethoscope and hypodermic syringes and needles as noted below.

Administering Medications

You will have to learn to give your sheep medications usually either by injection or orally. I'll describe how to give medicines to your sheep, but do get your vet or another expert to show you how at first hand.

There are two common sorts of injections, named after the injection site. The first of these is the intramuscular (IM) injection, so called because it is given in the flesh or muscle. I use a one-inch, 16- or 18-gauge (ga) needle mounted on a disposable syringe. For sheep, syringes with a capacity of 3 cc and 12 cc will handle most needs. The disposable needles and syringes are much to be preferred over reusable types because few of us have suitable facilities for proper sterilization and storage. Note: disposable syringes with the needle already attached are usually cheaper than buying the two parts separately, so read catalogs carefully.

After the syringe is filled, the needle should be inserted into a heavy muscle in the neck, shoulder, or rear leg. Avoid hitting a bone or a major blood vessel such as those located under the neck. After the needle is in place, withdraw the plunger to see if blood appears. If so, try another site because you do not want to inject directly into a blood vessel. For market lambs the rear leg should not be used for any injection that will discolor the flesh or cause an abscess. The IM site is mostly used for antibiotics, for which flesh discoloration is generally not a problem.

The other common site is under the skin or subcutaneous (called subcute, SQ, or SC). Pinch a bit of skin between thumb and forefinger and lift to form a small tent shape. Insert the needle into the elevated skin and slide it under the skin. Do not make just a shallow IM, but try to get the needle between the skin and the muscle. The best place for a subcute injection in a sheep is above the ribs. If long wool makes that site awkward, use the bare patch of skin just rearward of the front leg. As with the IM injection, be sure to withdraw the plunger to make sure you don't get blood.

To fill the syringe for a single injection, just shake up the contents of the bottle and withdraw the needed dose. If you are going to give injections to a number of sheep, you should not reinsert the needle into the container of medication after you have used it because you can contaminate the remaining contents. The recommended way to refill a syringe is to use two needles. Leave one in the bottle top, and use the second one to enter the sheep. To refill the syringe take the needle off and insert the syringe into the needle that is in the bottle. Fill the syringe, then remove it, leaving the needle in the bottle. Reattach the injection needle and go to the next sheep. This method does not totally prevent contamination of the contents of the bottle, but it is much preferred to using a single needle. Before putting a needle into the sheep, hold the syringe with the needle up and flick the needle end of the syringe with a finger to get the air bubbles to the top, and squirt out any excess air.

You also need a bottle of epinephrine to give in case an animal goes into shock after an injection. Never give an injection without having the epinephrine handy. You may never need it, but when you need it, you need it immediately. Give IM or SC 0.5–1.0 ml of the 1:1000 solution per 100 pounds body weight. May be repeated at fifteen-minute intervals as needed.

Some medications are given orally as a liquid or drench, using a special syringe called a drench gun or drench syringe. The end of the tube is inserted into the sheep's cheek alongside the molars, and the dose is given by pushing a plunger or squeezing a handle. The sheep's head should be held level, not elevated, so as to prevent the liquid from getting into the lungs.

Other oral medications are given in the form of big pills called boluses. The easiest way to give a bolus is to use a pair of forceps made for that purpose. The bolus should be placed over the hump of the tongue and the mouth closed. If the sheep spits it out, try again. If you have a helper, so much the better. The helper can straddle the sheep's neck, facing in the same direction as the sheep. The helper pushes his forefingers into the cheeks to force the jaws open. The sheep can't close her mouth without biting her own cheeks. After the bolus is behind the tongue the forceps and fingers can be removed. You will find it also helps to coat larger boluses with mineral oil to help them slide down.

Let me emphasize that you should get all of this demonstrated to you by an experienced person. There is no substitute for seeing something actually done correctly before you attempt it yourself.

If you are administering a drench or injection to a large number of sheep in a chute, it is important to do it in the right way to save your back and use the sheep's natural instincts. Do not lean over the side, because that will twist your back, and the sheep will back away or jump. Instead, get in the chute with the sheep, straddling them, and work from the back to the front. The sheep will generally back into you and between your legs and you can hold each under the chin as you work. If you try this from the front, the sheep sees you coming and will try to flee—right through you. You may get bruised or dumped and some sheep may miss their medication in the confusion.

Internal Parasites

Internal parasites—mainly worms—can be a big problem during the rebuilding time because there is a sharp rise in the activity of worms at lambing and during lactation. With timely worming, the problem is greatly diminished. Otherwise the ewes will come out of lactation with a heavy load of worms, and they will proceed to spread their eggs with their feces. The life cycle of sheep roundworms begins when the oocysts (eggs) are deposited on the ground. They then hatch, go through several larval stages, and attach themselves to plants that will be ingested by a sheep, starting the cycle all over again. Interruption of this cycle is the key to control.

Worming around lambing time (which will be discussed in the chapters on late gestation and lambing) is the most satisfactory approach for reducing the worms in the ewes. If this has not been done, an attempt should be made to disrupt the life cycle of the worms. If the ewes are not reinfected, their worm load will remain low for weeks.

Not to confuse the issue, but the biggest reservoir of worms on the farm is in the pastures rather than in the sheep. Larvae will hatch from eggs that survived the winter, and last year's pasture will be rich with newly hatched larvae as soon as the weather warms up. Those larvae will die off within three weeks, or if the pasture can be cut for hay, it will be relatively clean of worm larvae.

It is well to realize that there is no such thing as a worm-free pasture if sheep have been on it within the previous couple of years. Cold, dryness, and sunshine are the enemies of the worm larvae and the allies of the shepherd and sheep. A winter sun on bare pastures is an effective worm larvae killer, but eggs survive. Likewise, sunny, dry days in late summer when the grass is short are tough on larvae. Neither condition will destroy all of the larvae because the microclimate at ground level may be humid and pleasant to the worms, although it feels dry and dusty at the level of a farmer's face.

A useful strategy is to keep the ewes off pasture during lactation, supporting them on hay and grain. That can maximize milk production and prevent both reinfection of the ewes and, more important, infection of the pasture and the lambs.

With a pasture system, the worming strategy is one that minimizes worm larvae in the pasture. The cycle of egg to the next crop of eggs is about three weeks for common worm species. If pastured sheep are wormed at lambing, then at three weeks and at six weeks, that will help to reduce the number of larvae in the pasture. If possible, after worming, sheep should be held off pasture for twelve to twenty-four hours to prevent dropping of live eggs in the pasture. After the series of wormings, the pasture load will be reduced and also the ewe's immunity to worms will have recovered from a low point at lambing. In warm climates, an additional one or two wormings at three-week intervals can be useful.

Worminess in sheep will make its presence apparent by diarrhea, weight loss, unthriftiness, depression, loss of appetite, and other indications of poor health. In serious infestations, the skin beneath the ewe's eyelids (pull down the lower lid) and gums may be pale from anemia, and there may be swellings under the jaw (bottle jaw). If anemia is so severe that you find pale eyelids and gums, the ewe should be given 2 ml of iron dextran (a pig medication) to aid recovery. If you are butchering an animal for home consumption, the digestive tract can also be examined by a veterinarian. If worms are found, the sheep flock can be wormed and kept in a drylot or turned into a worm-free pasture or field—if you have one.

As far as what wormer (anthelmintic) to use, personal preference and custom often dictate a choice. There are a number of wormers on the market that are approved for use with sheep. The common ones are the benzimidazoles (thiabendazole, albendazole), levamisole (Tramisol, Levasole, Ripercol), and ivermectin (Ivomec), Most are for oral drenching, although levamisole is available as tablets too. Levamisole and ivermectin are available as injectables for cattle, but those forms are not approved for sheep. Read about drug approval in appendix 3.

When using a benzimidazole or ivermectin drench or paste, hold the ewes off feed for twelve to twenty-four hours before giving the oral dose to increase absorption. Do not do this with ewes in late gestation.

Some shepherds and veterinarians use wormers intended for horses. These are not approved for use with sheep, which does not mean that they are not safe and effective. I have never used any of these myself.

Most researchers today agree that tapeworms have little if any adverse effect on sheep or lambs. We stopped treating for tapeworms and found no ill effects in our lambs for almost two decades after stopping.

There is one tapeworm called *Echinococcus granulosis* that is a cause for some attention. The life cycle of this species includes a stage in which dogs are the intermediate host. In ruminants and in humans a larval stage lives in the liver, lungs, and other internal organs, where the worms can do great damage—even causing death. These pests can be controlled by interrupting their life cycle. Never feed raw sheep offal to dogs or let carcasses lie around where dogs can get to them. This is another good reason to keep stray dogs off your property. Pet dogs or farm dogs can be checked by fecal examination and wormed until free of *E. granulosis*. This worm is an especially serious risk if there are infants around who play with the dogs. Children may ingest the tiny eggs or egg-bearing worm segments when they put their fingers in their mouths. This is fortunately not a common problem, but be alert, and keep your dog wormed.

Resistance

Worms develop resistance to the wormers if they are used improperly. There are several practices that will reduce build-up of resistance. Perhaps the most important practice is not to use anthelmintics more often than necessary. Do not treat until clinical signs appear or fecal examinations show heavy infestation. The worming every three weeks mentioned above has its downside, because each worming increases the proportion of resistant worms in the total worm population. Never worm more often than every three weeks.

Wormers should be switched from one group to another every two to three years, or when one group becomes ineffective.

Dosage should always be at the recommended level. A safe practice is to weigh the largest sheep in the flock, and dose at that rate for all the ewes, unless you are able to weigh each ewe. Err on the side of too much anthelmintic.

The is a big Catch-22 factor at work here. After worming, one should not return sheep to the same pasture full of larvae, because the sheep will quickly become reinfested. However, one also should not move freshly wormed sheep onto a truly wormfree pasture such as a newly planted area, or one that has not had sheep on it for many years. The reason for that is that the freshly treated ewe will contain almost entirely wormer-resistant worms, and will infest the pasture with eggs from nothing but those resistant worms from that time forward.

For those who lamb in spring, a better strategy is to hold lactating ewes off pasture until the old crop of larvae has mostly died, then give the ewes pasture access after weaning. Then the pasture worm population with be a mix of resistant and nonresistant worms. If that is not possible, grazing of a pasture by nonlactating ewes, or even cattle, before turning the sheep in will help to reduce the total number of larvae.

External Parasites

The flock should be examined for external parasites. Confine the sheep in a small area, or use a sorting chute if you have one. Part the wool and look at the skin carefully. Sheep keds, often incorrectly called sheep ticks, are easy to spot. They are about the size of a housefly and are found mostly on the sheep's flanks and hips, although a badly infested animal may have them everywhere. Keds have six legs whereas real ticks have eight.

The other pest to look for is lice, and they are not as easy to spot. They are about half the size of a pinhead and the common species is a pale buff color that does not stand out well against the skin color of a sheep. The best way to check for lice is with a $10\times$ magnifying lens that can be purchased at a college bookstore or a good optical shop. Examine the sheep's skin or, better yet, scrape the skin with a pocket knife and put the scrapings on a piece of dark paper or cloth, and use the lens. If you are in doubt, get a veterinarian to examine the flock for you.

A shorn sheep is easier to treat than a wooled one. Shearing not only readily allows any pesticides to get onto the sheep's skin, but lots of the eggs, pupae, and adult keds and lice are removed along with the wool. This is particularly true of lice in summer because they glue their eggs to the wool fibers well away from the skin in hot weather.

The shorn sheep can be treated with a number of pesticides. The rebuilding time is especially suitable because there is no chance of the poisons injuring a fetus or nursing lamb. Powders can be used on shorn sheep, and we have had good results from malathion louse powders and even with garden rotenone dust. You will find that most of the pesticide powders make no mention of sheep on the label, so seek advice from a veterinarian or a sheep raiser.

If the sheep cannot be shorn for some reason, there are other options for pest control. The traditional treatment was to dip the sheep, immersing them entirely in a trough or pool filled with the sheep-dip material. Usually one or more people are stationed with special crooks to push the heads under briefly. Most of us do not have facilities for dipping.

The alternate method is to spray the sheep. Check with your vet for recommendations and availability of these materials, because what is approved and what is not seems to change on almost a monthly basis. There is risk to both the sheep and the shepherd with these potent chemicals, so consult experts and take proper precautions. Few pesticides are labeled for use with sheep because the relatively small number of sheep in the United States does not justify the expense involved to the manufacturer in getting formal approval.

Some people say that one can confine the sheep and apply spray solutions with a garden sprinkling can. I have never tried this because I couldn't imagine that long-wooled sheep could be completely soaked—a necessary element in pest control. A good way is to use a sprayer. You need a pump, sprayer wand and associated hoses, and a big drum to hold the spray mixture. You can use a pump powered by a tractor or get a more expensive self-powered type. The sheep can be moved through a chute with open sides and wetted completely on both sides and over the back by the jet. A suitable chute arrangement can be assembled from a few hog panels and steel fence posts for temporary use. With such spraying, be careful not to be so vigorous as to partly felt or cott the wool.

Another option is to use a systemic poison, meaning a chemical that is absorbed by the sheep so that when a ked or louse sucks a meal of blood it gets a lethal dose of the pesticide. A systemic poison is applied by pouring a

Sorting pens and chutes such as this commercially available British type make the shepherd's job a lot easier and faster.

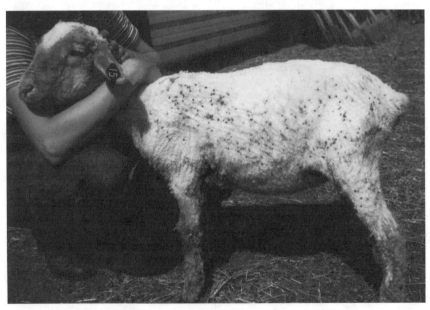

This yearling ewe weighed only eighty pounds—far less than her flock mates—because of this heavy infestation of sheep keds. (This ewe was not from my flock.)

Treating wooled sheep for external parasites can be done in a temporary chute using a tractor-powered sprayer rig.

diluted mixture down the sheep's back, by applying a paste, or by placing a small amount of a concentrated material on the sheep's skin. Dosage to the sheep must be carefully regulated, and the shepherd must take precautions to avoid contact with them. They are readily absorbed through the skin, so rubber gloves should be worn and contaminated clothing washed immediately after use. Ectrin is a pour-on approved for sheep.

It is important to remember that these chemicals may kill the sheep along with the lice and keds if they are not used correctly, so check with local experts. Unhappily, most of the pesticides are not approved for use with sheep by the Environmental Protection Agency, so the sheep raiser is left without official guidelines. Ask your vet, and take care with unapproved versions.

Dentistry

During the rebuilding period and before pastures begin to get a little sparse, the sheep's teeth ought to be examined to see if they are badly worn or missing. This ovine dental examination is called mouthing, and it is a simple thing to do. Dump the sheep on her rump or confine her to a crush or chair. To dump a sheep I hold her with my left hand under her chin, her left flank against my leg, and my right hand grasping her right rear leg in the wool-free area (the "leg pit"). Push down and back with your right hand as the sheep's head is turned back and to the right, and most sheep will sit down without a fight.

If sheep refuse to cooperate, grab them around the chest under the front legs. As they try to run away from you they will walk into an upright position on their rear legs. Then just sit them down. Don't you lift them up—let them do it. You can also back them into a 15"–16" junk tire or commercial sheep chair.

Once you have them controlled, lift each sheep's lips and look at the teeth. If a lot of them are missing, you should consider culling the ewe because she won't be able to compete on pastures that are cropped short. You can provide such a ewe with extra feed if you want to keep her. If the sheep has only one or two incisors left, it will do better if the remaining teeth are pulled; you can do this using a pair of pliers. Then the ewe can gum her grass if it is long enough for her to get a grip. Don't stop there; look at the molars too. If the molars are badly worn, there isn't much one can do except cull.

Bloat

Bloat may occur when sheep are turned into a new pasture, especially alfalfa. Bloat can also be brought on by stressful incidents such as being chased by a dog. I have even seen it happen when a helper unfamiliar to the flock was catching some sheep for me to shear. Mostly though, it is caused by overeating on fresh, green feed, especially alfalfa or clover. Gases become trapped in the rumen, usually in a foamy mass, the rumen swells, compressing the lungs, and the sheep may die from suffocation. Prevention is the key in this situation. Sheep turned into a new pasture that contains a lot of alfalfa should first be fed their fill of dry hay to limit their intake and then be allowed only a short time in the new pasture. As little as fifteen minutes the first day, thirty the next, and so forth might be appropriate.

Treatment of bloat is difficult, especially if a lot of sheep bloat at about the same time, because many will die while you are still treating the first ones. The traditional treatment is to cut a hole in the rumen to let the gases escape; this is called sticking the sheep, and there is a dramatic description of it (see facing page) in Hardy's *Far from the Madding Crowd*. There is a special tool called a trochar made for this purpose; it punctures the rumen and then maintains the opening so the gases can get out. Cutting a hole with a pocket knife is seldom effective because the hole doesn't stay open.

Bathsheba, with a sad, bursting heart, looked at these primest specimens of her prime flock as they rolled there—

Swoln with wind and the rank mist they drew.

Many of them foamed at the mouth, their breathing being quick and short, whilst the bodies of all were fearfully distended.

"O, what can I do, what can I do!" said Bathsheba, helplessly. "Sheep are such unfortunate animals!—there's always something happening to them! I never knew a flock pass a year without getting into some scrape or other."

"There's only one way of saving them," said Tall.

"What way? Tell me quick!"

"They must be pierced in the side with a thing made on purpose."

"Can you do it? Can I?"

"No, ma'am. We can't, nor you neither. It must be done in a particular spot. If ye go to the right or left but an inch you stab the ewe and kill her. Not even a shepherd can do it, as a rule."

"Then they must die," she said, in a resigned tone.

"Only one man in the neighbourhood knows the way," said Joseph, now just come up. "He could cure 'em all if he were here."

"Who is he? Let's get him!"

"Shepherd Oak," said Matthew. "Ah, he's a clever man in talents!" . . .

Gabriel was already among the turgid, prostrate forms. He had flung off his coat, rolled up his shirtsleeves, and taken from his pocket the instrument of salvation. It was a small tube or trochar, with a lance passing down the inside; and Gabriel began to use it with a dexterity that would have graced a hospital-surgeon. Passing his hand over the sheep's left flank, and selecting the proper point, he punctured the skin and rumen with the lance as it stood in the tube; then he suddenly withdrew the lance, retaining the tube in its place. A current of air rushed up the tube, forcible enough to have extinguished a candle held at the orifice.

It has been said that mere ease after torment is delight for a time; and the countenances of these poor creatures expressed it now. Forty-nine operations were successfully performed. Owing to the great hurry necessitated by the far-gone state of some of the flock, Gabriel missed his aim in one case, and in one only—striking wide of the mark, and inflicting a mortal blow at once upon the suffering ewe. Four had died; three recovered without an operation. The total number of sheep which had thus strayed and injured themselves so dangerously was fifty-seven.

—THOMAS HARDY, *Far from the Madding Crowd*

Surprised but helplessly under control is the only suitable description for a ewe in a well-designed cradle.

A better way to treat bloat if the time is available is to use a three-foot length of half-inch garden hose with the edges smoothed to prevent cuts. The hose can be lubricated and slipped down the sheep's throat into the rumen, permitting the sheep to swallow it as it goes. If the bloat is caused by blockage of the esophagus, the hose will shove the obstruction aside and the gas may escape with some speed through the hose (and the wise shepherd stands to one side). If nothing comes out, suck gently on the free end of the hose to draw a sample of the stomach contents into the hose. Place a thumb over the free end and withdraw the hose and sample. If there is a sudsy liquid in the hose, frothy bloat is the problem, and bloat treatment liquid is needed. Mix some bloat remedy, generically poloxalene (one brand is called Therabloat), according to directions on the bottle, reinsert the hose, and pour in an appropriate amount. Before pouring in the remedy, you might blow into the hose. If the sheep coughs you are in its lung, so try again. The hose can be turned gently to mix in the remedy and help break up the foam. You can also inject this directly into the rumen if you know what you are doing. Ask your

44

Sheep are easily restrained in this position—even if they do look pretty silly.

vet for instructions. If you have no frothy bloat remedy, a few tablespoons of vegetable oil and about a quart of warm water may help. Mineral oil can be used, but it does not have the antifoam properties of vegetable oils.

A bloated sheep is not only swollen—as the name suggests—but will look glassy eyed and have shallow breathing. Sometimes sheep eat so much that they look swollen. This condition is not fatal as long as the sheep has a chance to digest the rumen contents. Sometimes, though, a sheep will lie down when full and be unable to get up again. If a sheep lies down with her legs uphill, she won't be able to get up again with a full rumen, and she doesn't have sense enough to roll over and get up from the downhill side. Alternately, a sheep can roll into a ditch or depression and be unable to get to her feet. This is no laughing matter because the weight of the rumen can bear on the lungs and suffocate the sheep. If you see a sheep that cannot get up, run—don't walk— to her and set her up on her feet. She will stand there puffing and shaking her head for a few moments, then run off baa-ing to rejoin the flock as if nothing had happened.

45

Because a ewe will swallow a length of garden hose readily, the shepherd can easily check rumen contents for bloat or administer fluids or medication.

Escaping Sheep

Always glance at the fences and listen for distant bleats to check whether one of your beauties is stuck in a fence. Go over and get her out before she strangles herself. Grab her by one back leg and pull, and she'll pop out of the fence like a cork.

Sheep don't always get stuck in fences—sometimes they get through them. Some individuals are more talented at that than others, and they usually have what business schools call leadership ability—they take the whole flock with them. The solution to this problem is to fix the fences, or get rid of the offending individuals. Checking that the flock has not flown is a serious matter because you want to avoid offending your neighbors and letting the sheep bloat, but there are other concerns as well. Most insurance does not cover losses of animals that simply escaped through a weak fence or open gate. You are also liable for any damages caused by your sheep, and those damages might be considerable. If a driver swerved to avoid hitting your sheep and had a collision, you would be responsible. If bodily injury or death were involved, you could be out of the sheep business forever, especially near urban areas where judges and juries commonly tend to favor the city dweller over the rural.

Legs and Foot Problems

Look for animals with a limp and check any to see what the problem is. It may just be a rock jammed in the hoof or a pulled muscle. If the leg is seriously damaged, take the animal from the flock to recover or it will painfully follow the flock everywhere, in which case healing will be delayed and the condition may worsen. Put it in a small pen with another sheep for company, and let it mend.

If the hoof is odorous and has soft places the problem is probably foot rot. The affected sheep should be isolated because the disease is highly contagious. Treatment consists of severe trimming of the hoof to remove dead tissue and to expose the actively infected portions to the air. The hoof can be treated topically (i.e., at the site) with a 5 percent Terramycin ointment.

If possible, the whole flock should be treated by using a footbath of 10 percent zinc sulfate in water. Ideally all hooves should be trimmed before the footbath is used, and each sheep should be forced to use the footbath several times a day or to stand in it for an hour. Many veterinarians also recommend giving zinc orally at a rate of a half gram of zinc sulfate per head per day in the feed, or feeding a commercial product containing zinc methionine called Zincpro at the rate of one ounce per day for 100 to 130 head of sheep. It makes common sense to undertake any of these treatments with the advice of your veterinarian.

Foot rot in sheep is caused by two bacteria called *Bacteroides nodosa* and *Fusobacterium necrophorum* that exist synergistically, that is, in a cooperative relationship. They are both anaerobes (they live in the absence of air), which explains why trimming of the hooves is so important. *F. necrophorum* is a normal resident of sheep's digestive tracts and is usually present in barnyards and pastures. If *B. nodosa* is transmitted by an infected sheep it invades the soft tissues of the foot. Then the normally harmless *F. necrophorum* follows and causes an intense inflammation and a limping sheep, often to the point where the animal will walk on its knees to lessen the pain.

As with so many health problems, prevention is the key. Foot rot is very difficult to cure. Don't be overeager to return "cured" individuals to the flock. Be sure to quarantine any new animals, and if you visit a farm where animals have foot rot, clean your boots thoroughly so you don't bring it home to plague your flock. Also, keep any area formerly occupied by infected sheep free of any sheep for a week before using it for sheep again.

There is a vaccine available for foot rot that has had some success. It is certainly worth trying if your flock has a recurring problem.

More rarely, *F. necrophorum* acts as a synergist with *Corynebacterium pyogenis,* whose species name means pus former. The two bacteria enter the region just above the hoof and cause inflammation and swelling. Treatment consists of paring the hoof and draining the pus, plus topical application of penicillin as well as IM injections of penicillin. Have your veterinarian do this for you the first time so you can learn the proper procedure.

Flies and Health

Flies can cause problems in summer, apart from just annoying the animals. Not that the annoyance factor isn't important, because it is. Older sheep huddle together to protect one another from flies, and lambs go off feed if flies are present in large numbers. Barnyard sanitation is the only way to control flies, at least partially.

Flies will sometimes lay eggs in wet or dirty parts of a sheep's fleece or in skin folds that retain moisture. The larvae—maggots—cause irritation and death of tissue, and they live on the dead tissue and exudates from the wound. The maggots enlarge the cavities they occupy to the point that the sheep can die.

Control of fly strike—as this condition is called—consists of not giving the flies suitable places to lay their eggs. Tail docking and trimming of wool around the anus and vulva are useful preventives. Already infested sheep can be treated by first clipping off wool with scissors or shears, then applying a suitable insecticide that penetrates and reaches the maggots. A repellent should also be used to prevent reinfestation. If the maggots are deep into the tissue, a hydrogen-peroxide solution can be poured into the crevice. The foaming action will float out maggots as it cleans the wound.

One species of fly, *Oestrus ovis,* lays larvae near the nostrils of sheep. The larvae crawl into the nasal passages and sinus cavities to grow, causing considerable irritation. The larvae, called nasal bots, cause a sinusitis with a persistent mucous discharge from the nostrils. The adult flies also annoy the sheep greatly by crawling around their nostrils. Affected sheep hold their heads close to the ground and stamp their front feet, sometimes seeking cool, shady places to escape the flies. Bots are an annoyance rather than a serious problem, but can cause sheep to lose appetite and optimum health. Insect repellents sprayed around the nose area are of some help.

Eye Problems

Eye infections, commonly lumped under the general term pinkeye, can occur any time of year but are most prevalent in warm weather, when flies are active to serve as carriers (vectors) of the disease-causing organisms. Pinkeye in sheep is usually caused by one or more microorganisms in the groups called mycoplasma, chlamydia, or bacteria. Dust, wind, and strong sunlight are contributing factors. The eye or eyes are runny with tears, the membranes get red and swollen, and as the problem advances the eyeball gets a bluish haziness and ulcers form on the eyeball, which may protrude or even rupture. Both eyes are usually affected.

Affected animals should be kept out of strong sunlight, and topical salves, ointments, or powders should be applied to the eyeball. Eye drops of a homemade 5 percent solution of zinc sulfate in distilled water are effective for some pinkeye. One producer likes Cloxacillin Benzathine (Orbenin), which is a teat treatment for cattle—ask your veterinarian for approval. We found it very useful also to give an injection of an antibiotic, according to label instructions. Penicillin has been effective for one type of pinkeye we have encountered. For severe cases, an antibiotic can be injected into the eyelid, using a small-diameter short needle. Get your veterinarian to demonstrate this to you. An older treatment that is still favored by some veterinarians is to inject some sterile milk IM. The sterile milk can be prepared by dissolving a freshly opened packet of dehydrated milk in boiled, distilled water, and then boiling the solution for five more minutes. The solution should be cooled and used immediately. The milk protein stimulates the immune reaction in the eye and helps the sheep's body fight the infection. The same protein can, however, cause anaphylactic shock, a severe allergic reaction. The treatment for this rare event is an injection of epinephrine (adrenaline) according to label directions. Such a reaction can occur from any injection, so the shepherd should always have a bottle of epinephrine on hand just in case when giving an injection of anything at all.

In general, learn to spot a limping sheep or one with stiff legs. Look for any who are not eating eagerly or are lagging behind the flock. Check the one that hangs around the water trough more than normal. Be alert for a blind sheep or one with impaired eyesight. You may think that a blind sheep would

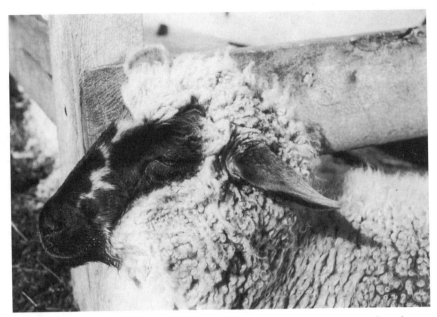

Eyes almost closed and drooping ears tell you this is a sick sheep. She recovered nicely after receiving some bloat medicine.

be simple to recognize, but you'd be amazed how well a blind one can get along by following the sound of the flock. Watch the ears too, for a sheep with drooping ears is probably a sick sheep that needs your help.

Polioencephalomalacia

Don't worry, you can call it PEM. PEM is a condition brought on when something goes wrong in the ewe's rumen where the "B vitamin" thiamin is produced. From a feed change or other cause, microorganisms that produce thiaminase, a chemical that destroys thiamin, become abundant and the sheep runs out of thiamin. This causes neurological effects—they look sort of drunk, become blind, and eventually go down and die. Affected animals do not have a fever. The treatment is thiamin. Get a bottle from your vet to have on hand. The affected animal should receive a big IM dose—10 cc or more. If PEM is the problem, recovery will be dramatic, with symptoms lessening in a matter of a few minutes. However, symptoms will recur in a few hours, so be prepared to treat repeatedly until the sheep fully recovers. If your vet does not have thiamin, use B-complex at higher dosage.

LAMBS

If the rebuilding time is a portion of the cycle when the ewes need minimal attention, that is more than made up for by the amount of care needed by the growing, weaned lambs—your big cash crop. It is important to change the lambs' environment as little as possible. For most of them the loss of their mother's companionship and built-in dairy bar is a severe change that will set back their growth and general health somewhat anyway, so all other factors should be kept the same.

Pasture or Feedlot?

A serious decision that has to be made is whether to allow the lambs onto a pasture. Pasture is a tempting alternative to a feedlot because grazing appears to be free, whereas the feed costs money out of pocket. Don't be fooled by this apparent difference, though. That pasture costs you money in a variety of ways. It is a fixed cost in the sense that you have money tied up in it and you pay taxes on it whether you use it or not. I'm not arguing that it be left idle, but it might make better sense to graze older sheep on it or to cut it for hay.

There are a lot of factors to consider in the feeding of lambs, but two stand out. First is the management of worm populations. If the lambs have been kept in a drylot so far, you can probably assume that their worm loads are pretty low. Remember that grass in a pasture is the natural habitat of the larval stage of sheep roundworms, so it makes little sense to put lambs there. Lambs compared to adult sheep have a very low resistance to worms, because the lambs' immune systems are not as well developed as the adults'.

The other factor is that lambs will not gain weight as fast on pasture as they will on a grain mixture. Pastures provide plenty of digestible feed with lots of protein, but the feed is low in energy compared to grain mixtures. We tried the full spectrum of lamb-raising schemes and concluded that the best way, for us at least, is to feed a high growth-rate mixture free choice and give only enough hay to keep the rumen working properly (about a quarter-pound a day per head). We use a mixture of grain and soybean meal similar to a creep feed (the name given to feed provided to lambs in the creep area; see chapter 7) to give about 14 percent protein with 0.5 percent salt and 2 percent limestone, as with the creep mixture. In keeping with lambs' conservatism, don't change their feed very much.

As mentioned in the chapter on lactation, you can give the lambs pelleted feeds from a manufacturer if you are willing to pay the higher cost. Make sure the feed is not too high in phosphorus because of the urinary calculi ("water belly") problem in rams and wethers. Using sufficient limestone in the grain ration is the best prevention. If urinary calculi are still a problem, addition of up to 2 percent ammonium chloride (an approved feed additive) to the feed will help. If a ram lamb has urinary calculi, sometimes it can be encouraged to pass out by snipping off the thread-like appendage at the end of the penis, then giving an oral drench of an ounce (28 g) of ammonium chloride dissolved in water, followed by daily drenches of 7 g of ammonium chloride dissolved in water. This treatment is usually ineffective, but will work often enough to be worth a try. Individuals can be treated by surgical amputation of the plugged urethra. It is usually more costly than it is worth to have a veterinarian do this, but the producer can learn the procedure if the incidence of water belly is high enough to require it.

Healthy lambs will gain about three-quarters of a pound to over a pound a day on a palatable grain-based ration and will convert the feed at a rate of about 3.3 to 4.0 pounds of feed for a pound of gain. You'll have to use local feed costs to decide whether intensive feeding of a grain mixture is the best economic move for you.

We found that our lambs do not finish well on hay and pasture alone. In regions of abundant pasture and low stocking rates, lambs can come off pasture in market condition in favorable years, though not always. We found that lambs fed hay after the pasture season gain very poorly and waste feed. Finishing without a concentrated feed is tough. The finishing feed need not be grain, of course, but it could be cull peas or beans, stale bread, dried potatoes, or a host of feeds peculiar to a given geographic area. Lambs can also be finished on crops planted especially for that purpose. See suggestions for flushing ewes in the next chapter for some suggested crops.

Many shepherds do not like to feed replacement ewe lambs on a straight grain diet because of damage to the rumen from the lack of roughage and a presumed shortening of their useful life and milking ability. We keep the replacements on a high grain diet until they get close to one hundred pounds, after which we increase the hay or pasture available to them.

Selecting Replacements

Most shepherds save some of the best ewe lambs for flock replacements or for selling as breeding stock. As a rule you will want to save lambs that come from proven ewes who are good milkers, have multiple births, and who have the traits you are trying to perpetuate in your flock. Some growers will say to save only twins and triplets. Others will select the ewe lambs that are the fastest growers. Selection on this basis may mean that more of the ewe lambs will settle their first year, but you may also be selecting for large mature size in the bargain. What the grower should be selecting for is productive ewes of a modest size who produce many pounds of good growing lambs. A big ewe may just eat more than a smaller one and yet not necessarily be more productive. Select ewes that come from the best mothers in terms of both prolificacy and milking ability. From that group you can then select individuals on the basis of their particular traits. You have to know which ewes are your most productive, which is why you need to keep good records for each ewe and her offspring over the years.

If you are saving rams for your own use or for sale, you probably should choose the fast growers. The ram has a big influence on the whole lamb crop, and you do want growthy lambs. A ram from a productive ewe and a large sire who is himself a fast grower is a good choice. Obviously, a big ram eats more and is harder to handle, but his lambs will probably earn you more money than a smaller ram. The ideal flock is made up of moderate-sized, productive ewes and big, growthy rams. Yes, another Catch-22—the big ram's genetics will become embedded in your flock if you save replacement ewes from him.

One cautionary tale—studies by J. J. Zenchak and others (*Applied Animal Ethology* 7 [1981]: 157–67) indicate that ram lambs raised in a male-only group show much less interest in ewes when they are used for breeding, so you should raise ram lambs that are to be used as breeders with a few ewes. The last thing a sheep breeder needs is a gay ram.

Health Problems

A few health problems apply only to fast-growing lambs. One of these (discussed in the chapter on lactation) is their susceptibility to *Clostridium perfringens*. The type-C strain causes scours in very young lambs but evidently is not a serious danger to older ones. The type-D variant is a danger almost throughout the lamb's active growth period, particularly when it is on highly nutritious rations. All lambs should have been vaccinated for both the C and D types at three to four months, or earlier if their mothers were not immunized in late gestation.

Lambs are also susceptible to infestations of internal parasites called coccidia, protozoans that inhabit the intestines of most sheep in small numbers. A least a dozen species are known in sheep. In range or pasture conditions, coccidia rarely cause problems, but in crowded lots outbreaks are frequent and extensive. The life cycle is similar to that of worms in that eggs are passed with feces. The eggs (oocysts) then sporulate and become infective in a day or two, after which they can cause infection of the lambs. The oocysts are highly resistant to heat, cold, and dryness, and they persist in lots and yards for long periods. Preventing reinfection by cleanliness is the best approach theoretically, but outbreaks happen even in seemingly spotless facilities.

The traditional treatment is to use sulfa drugs, which are generally given in the drinking water. Like most microorganisms, coccidia have developed resistance to drugs used against them, and the sulfas are pretty ineffective. Other coccidiostatic drugs that were originally developed for the poultry industry have been used on lambs with promising results. One of these, amprolium (Corid), is added to drinking water, and is reported to give control in some flocks. Two others, monensin (Rumensin) and lasalocid (Bovatec), are effective as feed additives. Lasalocid is approved for use with lambs, but monensin is not. Experiments have shown that about 10 g/ton monensin or 25 to 100 g/ton lasalocid in feed is effective in controlling coccidia (*American Journal of Veterinary Research* 42 [1981]: 57). Aureomycin, which is approved for use with sheep, fed at 10 g/ton of feed reduces coccidia oocyst output and improves weight gains. In studies in Canada by G. M. J. Horton it was found that monensin not only controlled coccidia but also improved feed conversion and allowed feeding a lower-protein feed mixture that gave gains equal to mixtures with 2 percent higher protein. Horton found 10 g/ton of monensin in feed to be optimal (*sheep!* 3, no. 1 [1981]: 18–19).

It is widely believed that a veterinarian can prescribe monensin to be added to feed. That is not true in the United States, because monensin is considered to be a feed additive rather than a drug. It is true in Canada.

For lambs on pasture it is more difficult to devise a preventive treatment for coccidiosis. Lasalocid has been mixed with salt at the rate of 0.75 percent (7.5 g/ton) and given free choice to pastured lambs. Treated lambs gained thirteen pounds more than untreated ones from weaning to market in a recent trial (*American Journal of Veterinary Research* 42 [1981]: 54).

Preventive action by treating ewes before lambing to reduce the spread to lambs is a wise move. Ewes can be treated in their feed or, in a pasture setting, decoquinate (Decox) in the mineral/salt mix for one month prior to lambing works well.

When any medication is given to a lamb, the animal should not be shipped for slaughter until enough time has elapsed for residues of the antibiotic or other chemical to disappear from the tissues. Approved drugs all have legal withdrawal times when used according to directions. The drug must be administered in the correct dose, by the correct method, in the correct site, all as directed on the label. Be sure to follow directions and do not ship until enough time has elapsed. Terramycin injectable lasts only 15 days, LA-200 28 days, and so forth. Read the label or ask your veterinarian if the label does not have this information or if it is unclear to you. You are responsible for the residues if they are found, and spot checks are made. Wormers and pesticides also have withdrawal times set by regulation. You can do pretty much anything you want with a breeding animal because it is not going to slaughter and nobody will ever know, but follow the rules with lambs or you may have a representative from FDA or a state agency knocking on your door. Remember that if you treat a lactating ewe the lamb will get a dose too. See appendix 3 for more details.

Flock Evaluation

At some point during the rebuilding period it is a good idea to get out the records from lambing and weaning and evaluate the ewe flock for possible culling. Those that didn't give you any lambs are prime candidates for shipping. If a ewe bred as a lamb didn't settle, you may well opt to give her another chance, but it is unwise to keep an older ewe who missed. She is unlikely ever to settle again. There is one exception to this rule. If a ewe has a sufficiently valuable fleece she might be retained. We have a few who produce fleeces that sell for ten dollars a pound, so they would pay their way even without lambs.

The ewes with broken mouths (i.e., missing lots of teeth) should go unless they are special for some reason, and I'd put the fence jumpers and other escape artists on the list too. Go through the records and seek out the ones with low lambing percentages. In today's market a ewe that produces only a single lamb is about a break-even proposition, and why should you go to all that work for a net profit of zero?

If you weighed your lambs frequently, you'll be able to identify the ewes that grow their lambs out well and those that don't. Any ewe that produces slow-growing, weak lambs is a liability to your flock and should be culled. Even a ewe that gives twins and triplets is not of much use if she cannot raise them. If a ewe has lost half of her bag to mastitis, she won't pull her weight. There are always exceptions, of course; we have a ewe with half a bag who always twins and raises both lambs.

If your flock is small, you may have some favorites that you are reluctant to send off for slaughter. At the very least you may not be very objective about culling if you have an emotional attachment to some of the animals. To prevent this pitfall you can rank your ewes numerically. Use a computer program if you have one available, or do some longhand arithmetic for a small flock. Decide on some categories that you think are important such as prolificacy, lamb-raising ability, general health, fleece quality, genetic background, or other meaningful characteristics. Then assign a percentage of importance value to each characteristic. As an example, let's say you broke down the list above as: prolificacy, 30; lamb raising, 30; health, 10; fleece, 10; genetics, 20. Then look at your records, and in the context of your flock, give a ranking of 0 to 3 for each category, with 0 meaning poor and 3 excellent. Multiply each ranking by the percentage assigned to each category and add up the total, which will range from 0 to 300. In the above example, if you ranked a ewe 2, 3, 2, 1, and 3 respectively, the arithmetic would be $(2 \times 30) + (3 \times 30) + (2 \times 10) + (1 \times 10) + (3 \times 20) = 240$. If you prefer percentages, divide the sum by 3 to get 80 percent. You may find that some of your favorites end up with distressingly low rankings and should be culled. If you haven't the heart to ship them, get a sheep-raiser friend to come over and talk you into it. It isn't in your long-term interest to keep them.

2

FLUSHING

The couple of weeks or so prior to breeding is the time to prepare ewes and rams for their job of manufacturing lambs. The ewes will have recuperated from lactation, and the rams will have idled away their time under a tree until they are ready to perform. If rebuilding was done right, the ewes and rams will come into flushing in fit condition, neither too thin nor too plump, but perhaps leaning a little to the slender side of the scale. Living things, from the tiniest microorganisms to complex higher forms, respond to the availability of nutrients in their environment by altering their reproductive rates. There is a natural feedback system in the biological world that regulates numbers according to their food supply. In the higher animals, low levels of nutrition are reflected in a low sperm count and libido in males, and low rates of ovulation and a reduced number of heat or estrus periods in females. Conversely, all of these increase when food is abundant.

Flushing is an old shepherd's term for increasing the feed to the flock in order to get them ready for breeding. The idea is to put the sheep on a rising nutritional plane in hopes of causing more ova to be shed at each heat and starting the estrus cycles sooner. It is a matter of making the ewes' insides think that good times are coming.

Flushing is most effective with slender, older ewes (two-year-olds and older) during the early or late parts of the normal fall breeding season. It has less effect on lambs and yearlings and in the middle of the usual breeding season. Also, a flock that is already in adequate condition may not benefit from flushing and may instead become too fat.

The rams should have been maintained on a pretty good diet because they build their sexual apparatus all summer, with maximum testicle size usually attained in July in the Northern Hemisphere. Just be sure they have had enough exercise to be fit and ready to go.

After a long period of relative quiet, the flushing stage is a time when the shepherd as well as the sheep begins to feel the oncoming excitement of the breeding season. A few ewes will start to cycle early and stand around bawling. The rams will begin to get off their duffs and look longingly through fences and gates at the ewes. The shepherd, who has been pretty relaxed during the building time, has a good feeling knowing that the pattern of breeding, gestation, and lambing is about to start again. After all the necessary nonsense of haying and taking care of crops, it is great fun to sit down and decide which ram to breed to which ewe. It's fun to get back to being a shepherd again.

NUTRITION

At the outset, remember to change feed slowly when an alteration in diet is made. This is especially important when a change is made to a more concentrated feed, especially one rich in carbohydrate energy. An increase in a bulky feed such as hay won't cause much of a problem because it is relatively low in quickly digestible energy. A sudden boost in a high-energy food such as corn can create trouble.

Rumen Changes
The thing to keep in mind is that the microorganism flora in the rumen of a sheep is a complex community made up of many different species and strains, each of which is adapted to a certain kind of food. When the ewe's diet is altered, the relative numbers of the various microorganisms are no

longer appropriate to the task of digestion. Some types will have to increase in number while others diminish. The whole character of the rumen population must change, and that takes time. The careful shepherd changes the diet by steps over a period of a week or so to get to the final high nutrition flushing fare.

Failure to allow for slow change can have highly adverse effects. A sudden increase in carbohydrates can cause a condition called lactic acidosis. Corn or other cereal grain is the usual cause, although wheat is even more dangerous, because it is so easily digested that fermentation in the rumen is very rapid. Wheat should never be more than half the grain fed, and generally less than that. Acidosis also can be engendered by green corn, potatoes, sugar beets or cane, forage turnips, cabbage family crops, fallen fruits, or a host of other culprits. The change from a high-roughage, low-energy diet such as pasture or hay to a low-roughage, high-energy one creates the problem. With the sudden availability of carbohydrates the starch-converting species increase in numbers very rapidly and dominate the rumen flora, producing lactic acid, but other microorganisms that utilize the lactic acid do not increase sufficiently rapidly to convert the lactic acid to a useful and harmless form.

Slow changes in feed prevent all of this. Some ewes, especially older ones, can get acidosis even with slow changes, especially if they are aggressive eaters. They will droop their ears, get diarrhea, and look very sick. Give them a drench of a couple tablespoons (up to a quarter cup) of baking soda (sodium bicarbonate) dissolved in a little water, and they'll snap right out of it. Some producers give mineral oil to empty the rumen and to slow absorption of the lactic acid. A turkey baster is handy for all of the above dosing.

If undigested carbohydrate passes from the stomachs into the intestine, it can cause a rapid increase in the numbers of the bacterium *Clostridium perfringens* type C, the same bacterium that causes some scours in very young lambs. *Cl. perfringens* ferments sugars, producing gas and acid as well as a toxin that can kill the sheep. This disease is called "struck" in England, and is supposedly rare in North America, though we had it in our flock before we started vaccinating against the type C variety. It mostly affects yearlings and two-year-olds. If the intestine has not been damaged beyond repair, affected animals can be saved by injections of large amounts (10 to 40 cc) of appropriate antitoxin. The best approach is preventive: vaccination and slow feed changes.

Legumes

As long as the changes are made relatively slowly, almost any feed can be used for flushing. There is one feed to avoid, however, and that is clover, especially red clover: clover and some of the other legumes, such as alfalfa, contain phytoestrogens—chemical compounds that can inhibit estrus and ovulation. A ewe fed large amounts of red clover won't come into heat at all, so if you feed red clover, do it at some time other than flushing or breeding.

No pasture is 100 percent red clover, of course. Brian Magee at Cornell suggests a bioassay of the pasture using wethers. Check the wether's teat length, then turn them out on suspect pasture for two to three weeks. Remove them and monitor teat length for a week or two. If it more than doubles there is lots of phytoestrogen present, and one should keep ewes for breeding off that pasture.

One could, at least theoretically, feed red clover intentionally to hold off estrus until desired. I have not tried this, so I can't tell you how long the birth-control effect persists after the clover is withdrawn, but it might be a good way to crudely synchronize cycling of a ewe flock.

Other legumes do not seem to cause estrus problems. We have fed alfalfa hay through flushing with no ill effects, and even ewes flushed on standing birdsfoot trefoil have come into heat while eating it. I suppose to be on the safe side one could feed a nonlegume hay at this time if it was available.

Flushing on Pastures and Crops

By far the simplest flushing procedure is to turn the ewes into lush pasture if it is available. They will thrive and be ready for the rams. If seasonal or other considerations make ordinary pastures unsuitable, then the flock can be flushed on crops planted especially for the purpose.

Rape, a primitive cabbage, is a common flushing crop that is usually planted as a companion to oats, barley, or other small grain. The rape grows slowly under the canopy of stems and grain heads, then takes off rapidly after the grain is harvested.

Forage turnips, including the forage turnip × Chinese cabbage cross, Tyfon, can be planted in the same way to give a late summer or fall crop for grazing. If cheap seed can be located, any member of the cole (cabbage) family can be planted. The southern "green," collards, produces large quantities of palatable forage in any climate, or kale might be considered for the north because of its remarkable frost resistance. Choose your kale carefully, because some varieties in quantity are poisonous to sheep. Sheep feeding on any cole family plants should have access to iodized salt in their salt/mineral mix because the coles contain compounds that inhibit thyroid function and can produce goiter if sufficient iodine is not available.

Rather than planting a flushing crop as a companion to small grain, some shepherds plant a field or two especially for flushing. A good way is to mix a variety of seeds and dump them all into a seed drill and plant without even tilling the field first. The drill will get the seeds into the ground if it is moist enough, although you may till the field lightly if you wish. Try mixing oats, barley, wheat, millet, corn, rape, peas, soybeans, or any other similar seed you might have. A mixture of plants in such a planted pasture gives the sheep a rich variety of excellent nutrition, generally better than any the plants could provide separately because of their different root depths and the variations in the composition of their tissues. Because flushing is commonly a fall activity, it is wise to avoid sorghum and other plants that can cause hydrocyanic acid (prussic acid) poisoning if stressed by drought or frost.

Another possibility is to plant a fast-growing crop like Tyfon into crop residues such as stubble from corn or small grains. The Tyfon will be ready to graze about seven weeks from planting. Straw should not be tilled in, as it would use up nitrogen in decomposing, and Tyfon and other turniplike plants are big nitrogen consumers. A sheep-raising friend in Pennsylvania, Jim Barlow, seeds Tyfon after harvesting a sweet corn crop and uses it for fall grazing. Tyfon can also be sown by air about two weeks before grain harvest to give it a head start. The suppliers of Tyfon, Great Western Seed Company, P.O. Box 387, Albany, Oregon 97321, recommend sowing about five pounds to the acre, or a bit more if by air. Seed is also available from Lehle Seeds, 1102 S. Industrial Blvd., Round Rock, Texas 78681, sales@arabidopsis.com, www.arabidopsis.com/.

Common purple-top turnips and forage turnips also make a fine flushing crop. A planting of two to three pounds to the acre gives nice big roots as well as tops. Both turnips and Tyfon require about 100 pounds of nitrogen to the acre. Turnips don't compete well with weeds and are best planted following a clean crop. They are ready for grazing after about ninety days and are best suited to regions of mild winters where they won't get caught in frozen ground. Small roots of either Tyfon or turnips can lodge in sheep's feet, so be alert for that, especially in muddy ground. With a good crop of either plant you can expect to be able to graze about 180 sheep for two weeks on one acre.

An advantage of turnips over Tyfon is that they will store themselves, as it were, until the sheep are ready. Tyfon, on the other hand, should be grazed immediately on maturity, and it will grow back quickly if your climate allows it. Sheep have been known to refuse drought-stressed Tyfon, a fact you might consider if dry growing seasons are common in your area.

Trials of brassica forage crops in 1981 by Frank Schwulst at the Colby, Kansas, Agriculture Experiment Station resulted in the following grazing days per acre: turnips, 2,300; turnip rape, 3,400; fora rape, 3,400; and Tyfon, 2,400. Fora rape is a forage rape variety that grows as high as three feet and is thus quite resistant to loss by trampling, unlike some related crops. Other possibilities in the same broad group of plants include rutabagas and large radishes. The giant Japanese radishes, or daikon, would be an interesting possibility to try if an inexpensive source of seed could be located. I suppose the sheep would burp a lot though. For seed of any of the above try Advanta Seeds Pacific in Albany, Oregon, 800-288-7333 or White Lake, Wisconsin, 800-359-2480; Albert Lea Seed House, Albert Lea, Minnesota, 800-728-8450; Barenbrug USA, 302080 Old Highway 34, P.O. Box 239, Tangent, Oregon 97389, 541-926-5801 or 800-547-4101; and Welter Seed and Honey, Onslow, Iowa, 800-728-8450.

Some sorts of crop residues are suitable for flushing provided they are not just roughage such as corn stalks. Sugar beet tops are excellent, provided the sheep learn not to choke on them. Corn, pea, and bean residues make good feed if there is enough lost grain and not just stems; the same comment applies to small grain leavings. If the combine didn't get everything, and they never do, sheep will do a dandy cleanup job. If spilled grain is abundant, the sheep should be allowed access for only a short time each day—as little as ten or fifteen minutes—so they do not overeat (with the distressing results discussed earlier). In southern areas sugarcane stubble can be used.

Green chopped crops make good flushing feed too, with corn being especially good if it is chopped before maturity. With green chop the shepherd can go out each day and chop just enough for one feeding at a time. Sugarcane and alfalfa grass mixtures can be used in the same way, or a field of peas can be planted for chopping. Straight green chopped alfalfa can cause bloat, so caution is advised.

Unusual Feeds

In coastal areas don't overlook seaweed as a source of sheep feed. Feeder lambs are bought in Britain and fattened on French beaches to provide a supposedly particularly tasty product for Paris gourmets. It would be difficult to graze sheep on many U.S. beaches, but one could harvest large quantities of the giant algae with a dump rake and a loader at little cost. Sheep raisers near the coast should have some of the local seaweed analyzed for feed value or feed some on a trial basis. Shepherds near coastal government or military installations with long stretches of fenced-off beach should look into acquisition of grazing rights. You could become the only shepherd in the area who had to get a security clearance to graze your sheep. What the sand does to the sheep's teeth I have no idea.

In some areas you might be able to graze your sheep to control weeds in crops fields or to reduce noxious plants in grazing land. New Zealand farmers use sheep as four-legged herbicides to clean up fields as a normal part of the crop cycle. There are experiments with using sheep to rid grazing land of leafy spurge, poisonous to cattle but harmless and nutritious to sheep. Sheep are being used to rid city parks of kudzu in one Florida city.

In some areas spent grain from the fermentation process may be available from breweries, distilleries, or fuel alcohol production plants. Such grain is low in energy and is wet and heavy. Brewer's yeast could be available from the same sources. Either would have to be supplemented with a roughage and energy source, though the waste products themselves are very high in protein.

For energy sources to go along with high-protein feeds, use your imagination. Can you find a supplier of reject or spoiled fruit? Just be sure it hasn't reached the fermentation point, or you might have a bunch of woolly drunks to take care of. Do you live near a plant that processes potatoes? The peelings and odd bits left from making French fries are a fine energy source. Both fruit and potatoes are high in water content, so dry roughage would have to be fed along with them to head off diarrhea. Stale bakery goods are commonly a cheap source of dry, high energy feed and are available in huge quantities from large bakeries. Beware of feeding too much of the bakery goods, because they are very easily fermented in the rumen and acidosis can result; however, together with hay they are just fine (see Rumen Changes a few pages back).

Don't overlook leaves as sheep feed. Sheep raiser Gary Jones of Peabody, Kansas, fed leaves as the sole feed to his flock when a drought made hay scarce one year. He reports that leaves from osage orange tested almost as well nutritionally as high-quality alfalfa hay. He put a classified ad in a local paper that offered cash for bags of leaves delivered to his farm and was practically buried in the huge pile of leaf-filled garbage bags that suddenly materialized. A combination of leaves and grain might make just the flushing diet for your flock.

The commonest flushing fare is grain, especially corn, fed along with hay or pasture. Any grain can be used, of course, and you should let local prices and availability be your guide. We generally feed a pound of corn and three to four pounds of hay per ewe. Unless your hay is straight alfalfa, which is high in calcium, add some limestone to the grain at the rate of 1 to 1.5 percent (twenty to thirty pounds per ton) to balance the phosphorus in the grain. Limestone can also be added to a salt mix.

Be sure the ewes have plenty of water available. On reduced water, sheep will voluntarily limit their feed intake, and you want them to eat well. Make salt available to them to encourage water drinking. Most authorities recommend mixing about one half dicalcium phosphate with the salt or using a commercial sheep mineral mix. The latter is much more expensive.

Whatever the feed, be sure that it is one on which the ewes will gain. Use the NRC tables or try some feed computer software, then get out there and feel a few backs to see if they are gaining weight. Remember also that overly fat ewes will have low fertility and should not be put on a gaining diet at all.

For flushing, the environment can be a forty-acre field or a small feedlot. If you have no need to observe the ewes closely it doesn't really matter. We like to check on our ewes frequently during flushing, and we feed them their grain in a small, confined area so we can examine them with minimum difficulty.

It is also helpful to move the sheep around a bit to get them used to being herded again if they have been loose on pasture for some time. Feeding grain is a great training aid to reacquaint them with the concept of coming on call. They quickly learn to respond to a call of "Sheep!" or "Oh-oh-ovie!" when they are rewarded with some tasty corn or oats.

Prebreeding Activity

Teasing

Flushing is the time to start teasing if your management scheme includes this practice. Teasing is placing a sterilized ram with the ewes to aid in bringing them into heat but not getting them pregnant. The teaser is commonly fitted with a marking harness and crayon so the time of heat of each ewe can be noted from the crayon mark left on her rump by the teaser mounting her.

The idea behind teasing, apart from indicating the time a ewe is in heat, is that ewes produce fewer ova on the first heat than on subsequent ones, and more ova mean more twins and triplets. The sterile teaser ram brings the ewe into heat and marks her. The marked ewes can then be put with a fertile ram on the next or later heat cycle for breeding. We have found that our lambing percentage is greatly increased by teasing, and I recommend it highly to any producer who wants more lambs. We do not tease ewes being bred as lambs because we try to keep the number of multiple births from lambs at a minimum.

The teaser can also be a fertile ram wearing a harness, called a Ram Tam, that prevents him from entering the ewe. We have not used those harnesses on any of our rams, so I have no firsthand experience to report, but one does wonder about their cleanliness and the effect on the ram's libido. Another method is to put a ram in a pen or pasture next to the ewes so that he can be smelled by them but cannot get at them to breed them. This method is not always satisfactory because the ram may jump the fence or break through a gate or panel to get with the ewes—or even breed them through the fence.

Yet another method is to use a cryptorchid ram. Cryptorchism refers to the testicles' remaining in the body cavity instead of descending into the scrotum where they belong in a normal ram. This condition can be artificially created by pushing the testicles up into the body of a ram lamb and placing an Elastrator band where the empty scrotum attaches to the abdomen (as described in the castration section of the chapter on lambing). The crypt ram has all the normal hormones of an entire ram, but the sperm are killed or weakened by the high temperatures in the body cavity, so that theoretically the ram can tease the ewes without getting them pregnant. The only problem with cryptorchids is that they are not always completely sterile.

The most satisfactory method is to have one or more rams vasectomized. The veterinarian removes a section of each of the tubes that carry the sperm from the testicles (the vas deferens) and ties off the free ends. Sperm are then confined to the testicles and resorbed. The ram is a male in every respect except that his semen contains no sperm. Be sure that the vet doing the surgery removes a section of each vas. We had one vasectomy done in which the veterinarian merely tied the vas tubes. This worked the first year, but then the ram became fertile again somehow and bred almost the whole flock the next year when we thought we were using him as a teaser—disaster! In addition, be sure to have the vasectomy done sufficiently ahead of time so that sperm in storage weaken and die before the teaser is placed with the ewes. We have found that two months is long enough.

The teaser can be equipped in some way to mark the ewe when he mounts her. A harness that holds a marking crayon is the usual device. The awkwardness of a marking harness is another reason why I don't favor the use of yet another harness to prevent entry into the ewe. With all those straps and buckles, the ram looks like a sky diver who has lost his airplane. He might feel less interested in sex than in getting rid of his encumbrances. (See more about types of crayons and the like in the chapter on breeding.) You can also smear paint or grease every day over the ram's brisket, with which he marks the ewes. When a ewe is placed with a fertile ram, he can wear a different color crayon or paint so that the actual breeding date can be noted. Heat periods should be about fourteen to nineteen days apart, depending on the breed of the ewe, with sixteen days usual for most breeds. The shepherd should check the ewes daily for marks: another good reason for feeding in a confined area so the spookier ones can be examined almost as easily as the laid-back ones.

A teaser ram can also be used to advantage in management schemes attempting out-of-season breeding. Experiments at the University of Wisconsin have shown that the number of ewes coming into estrus out of regular season is increased by the continuous presence of a teaser ram, so a teaser can be useful more than just once a year. If you raise sheep partly for handspinner's wool, choose a ram with a valuable fleece, and he will pay his feed bill with wool alone. Otherwise, pick a lean, lanky fellow who won't eat too much, but who has lots of sex drive. Do remember, too, that the teaser may not be fertile, but he still has a ram's personality, so don't turn your back on him unless you enjoy bruises and broken bones, because many rams are aggressive by nature and will take any opportunity to butt the careless shepherd. Even seemingly tame rams will butt when the notion strikes them, and pets are often worse because they have little fear of people. I have had bruises and broken ribs from rams, and one of them put Teresa in bed for a couple of days, so don't take chances.

Handling

Even if your ewes are being flushed on pasture, you will still have to handle them a bit. The main thing to do is to shear both the ewes and rams, at least partially. The ewes can have the wool trimmed off around their vulva, or "crutched," to give the rams a clear shot as well as to give the shepherd a clear view to observe early signs of heat. For long-wooled sheep it can also be helpful to shear over the rump to provide a place that will show the crayon mark clearly. Some crayons don't mark well on long wool, especially in cool and wet conditions. The rams should have their bellies sheared, and in warm weather the wool clipped off their scrotums as well. You may prefer to shear them completely, especially if you do breeding in warm weather. Consider shearing any long-wooled breeds completely at this time too. With Lincolns, Cotswolds, and their crosses the wool is so long in a full year that the fleece can be lost to tangling and cotting of the ends. With these breeds a six-month fleece is long enough for most purposes.

If estrus synchronization is practiced, you should insert vaginal sponges or pessaries at this time. I'll discuss synchronization at greater length in chapter 3 but the idea is that chemicals in the sponges or pessaries keep the ewes from estrus as long as they are in place. They would be removed prior to placing a teaser ram with the ewes, or when the ewe was placed with a breeding ram, if teasing is not used.

Medical Care

During crutching of the ewes and shearing of the rams' bellies and scrotums, the hooves should be checked and trimmed; look for any signs of hoof disease in the process. Be sure to trim the hooves enough, but try to avoid bleeding, especially of the rear feet.

Ewes have to bear the weight of the ram on them, and he must have sound feet too. To trim feet you will need a pair of what the catalogs called hoof rot shears. (This is, to my mind, a pretty negative attitude because well-trimmed hooves are less subject to infection.) I personally prefer tree-pruning shears.

If you look at a sheep's hoof, you will see little growth ridges on the surface of the hoof. You should trim parallel to these growth lines. Trim off any ragged edges down to fresh hoof. Then trim the point of the hoof until the cut reaches the fresh hoof material that looks a little bit like a hunk of polyethylene plastic or paraffin. You may find it simpler to first cut a piece off the tip. You will not have to trim much off the heel unless it is rolled under badly. When you get close to making the hoof bleed, you will be able to see blood faintly through the translucent material of the sole of the hoof.

For trimming, I dump the sheep as for shearing, and with it controlled between my knees, I trim the back hooves first, then the front ones, so a freshly

Hoof trimming: side view of hoof before trimming. Note the growth lines on the hoof.

Side view of hoof after trimming.

Bottom view of untrimmed hoof with rolled-under edges.

Bottom view of hoof after trimming.

sharpened front hoof can't cut me while I'm trimming the rear ones. Some people prefer to tether the sheep to a post at the corner of a pen using a halter, then lift the hooves one at a time to trim them, holding the sheep against the side of the pen with legs and hip. Some sheep will stand docilely for such trimming.

You should not have to trim hooves more than twice a year if sheep are out on rocky or sandy pasture most of the time. Individuals that require more frequent trimming are candidates for getting out of the flock.

Check and treat the ewe's vulva for any signs of infection or cuts. A ewe with a pink and swollen vulva may simply be in heat, but the shepherd should be sure that infection is not the cause. Presence of a clear, sticky discharge is normal for a ewe in estrus, but if the discharge is cloudy, dark, or malodorous, treatment should be given. Consult your vet for recommendations.

Be sure also to check the ram's penis sheath (the prepuce) for obvious sores or scabs, and if these are present, treat according to your veterinarian's advice. Examine each ram's testicles while he is in a standing position. A tenderness or swelling of the tube at the bottom of the testicle (the epididymis) is a possible sign of epididymitis. Have your vet confirm your diagnosis and cull the ram if he really has epididymitis because there is no effective treatment. It is caused by an infection with *Brucella ovis,* and can spread from ram to ram.

Disease prevention is always better than attempting a later cure. Breeding is a suitable time to vaccinate for a number of diseases of sheep, but especially for vibriosis. (Other diseases of local importance may be of concern, and nearby producers and veterinarians should be consulted.) Vibriosis used to be confined to the Rocky Mountain states, but is now found throughout the United States and Canada.

This disease is caused by a bacterium, *Vibrio fetus* var. *intestinalis,* and results in weak lambs, stillbirths, and abortions. The disease is transmitted from carrier ewes who show no clinical signs, or from aborted fetuses and associated tissues and fluids. Abortions usually occur during the last trimester of pregnancy and can affect large numbers of a flock in so-called abortion storms, with losses of up to three-fourths of the lambs. The bacteria live in the intestinal tract of carrier ewes between outbreaks. Prevention is by vaccination up to thirty days before breeding. The vaccine is not the same as the one for vibriosis in cattle. New stock or replacements should have been vaccinated twice when introduced to the flock and revaccinated at this time. It is commonly believed that once a flock is vaccinated against vibrio, the producer is obligated to vaccinate every year because the vaccinated ewes become carriers of the disease. This is nonsense. A revaccination every few years, or even every year in endemic areas, is not out of order, however, because the disease can be transmitted by carrion-eating birds or by a pair of dirty boots.

Another abortion-causing organism is a chlamydia that causes enzootic abortion of ewes (EAE). Enzootic abortion was once confined to Idaho and a few other states, but with the transportation of sheep about the country, it is, like vibrio, now found throughout the country. Once again, vaccination is in order as a preventive measure. A combined enzootic abortion and vibriosis vaccine is available. There is no need to worry about EAE if it does not occur in the region where you live, but that's not likely.

If sore mouth is present in your flock, you might want to vaccinate the flock at this time. Sore mouth is caused by a virus that persists in soil and around barns. Sheep that have had sore mouth, either from an accidental infection or from vaccination, are not carriers, but scabs shed from their infection contain live viruses that can infect susceptible animals. If you have never had sore mouth on your farm, do not vaccinate, as this will introduce it. Your best bet is to do nothing until the disease appears.

If sore mouth is already present on your premises, you can give the ewes a vaccination now and perhaps a booster in late gestation to make them immune. The vaccine is a live virus in a powdered form that is mixed with a diluent to make a suspension that is applied with a brush. A place on the skin under a leg or in an ear is abraded (not enough to cause bleeding) with a tool supplied with the vaccine, and the vaccine brushed on with the bristles at the other end of the tool. Alternately, a bit of wool can be plucked from the breastbone of the sheep and the liquid applied there.

The sites can be checked in a week, and the ewes revaccinated if there was no "take," a condition indicated by the presence of a scab. Do not try to save the vaccine after rehydration because it loses effectiveness quickly. The vaccine is only about five to seven dollars for enough to vaccinate one hundred sheep or more, so don't be penny-wise. One precaution: the virus can affect humans and cause a disease called orf, so wear gloves and wash your hands thoroughly after using the vaccine.

When ewes overeat on grain there is always the problem of lactic acidosis (as mentioned earlier), but there can be a much more immediate health problem that is purely mechanical. An occasional ewe will eat so eagerly that she will get a mass of grain lodged in her esophagus that will block the passage of air through the trachea to the lungs. The result is rapid death. Lightweight grains such as oats are especially bad in this regard because they can even be inhaled into the trachea itself. The symptom to watch for is a sheep coughing violently or trying to cough. It will try so hard that the spasm will lift its front feet off the ground, sometimes to the point that the sheep will fall over backward. The shepherd must act quickly. Have a short length of garden hose handy, one-half inch for adults and three-eighths of an inch for lambs, with the ends smoothed by a little sanding or filing. Lubricate the hose with a little mineral oil (or water if that's all you have) and ease it down the sheep's throat to try to clear the esophagus by pushing the plug of grain into the stomach. You may be able to feel the grain plug with your fingers from the outside of the throat and manipulate it from the outside as the hose moves along.

A physician friend of mine reminded me that the well-known Heimlich maneuver would be useful to dislodge a grain plug in the trachea or at the junction between the trachea and esophagus, and he pointed out that Dr. Heimlich first tried the technique on animals, including sheep. The idea of this technique is to cause a burst of air to be expelled from the lungs so as to dislodge the offending object. You will have to apply a sharp, forceful pressure to the soft part of the sheep's "stomach" between the navel and the rib cage, perhaps with fists or a knee, but don't get so violent that you injure the sheep with the treatment.

The only other observation that is really important at this time is to keep an eye on the ram to see that he is doing his job of teasing and is actually mounting the ewes. If he acts tired or has a limp, take him out for part of each day so he can get a rest from his vital chores.

EVALUATION

The flushing period is your last chance to cull unproductive ewes, so don't delay. There is no sense feeding a ewe all through breeding and gestation if she is not going to give you lots of lambs.

This is also the time to think about which ewes should be bred to which rams in the interest of both long-term genetic changes in the flock and top lamb production. It is too late to change your mind once the ewe is bred by a fertile ram.

Flushing is also the time to make a judgment about ewe lambs if you intend to keep any of them as breeding stock. Each sheep raiser will have somewhat different specific criteria to use in judging a ewe lamb's suitability as a breeder, but some general rules apply to most situations. Poor gainers should be shipped when marketable because they will generally not prove to be productive ewes. They may well be slow growers because they do not eat enough, and that means they won't eat enough to be good milkers either. If a small lamb is a triplet or a quad, then the small size may be acceptable, provided you find it economical to hold her over for breeding as a yearling.

Ewes bred as lambs should weigh at least one hundred pounds when bred. If bred at lighter weights, they commonly will settle, but will produce poor lambs, give little milk, and probably will be permanently stunted by the premature breeding. The lambs that make the weight should be evaluated for body conformation, wool quality and type, and how well they fit a flock or breed standard. Big, healthy lambs that don't fit your flock for some reason should be sold as breeding stock if possible, because you'll get a better price for them than you would if you sold them as slaughter animals. You'll also be making quality breeders available to others.

One management technique is just to breed all the ewes that meet the minimum weight standard and then ship the ones that don't get pregnant. They can be pregnancy-checked as described in chapter 4 and the pregnant ones retained in the flock, adding their tendency for early sexual maturity to the flock gene pool. The open ones will still be young enough to go to market as lambs at a higher price than you'd get for them as yearlings, and you save the cost of several months' costly feed.

On the other hand, you may prefer to retain all suitable ewe lambs to be bred as yearlings. In times past, very few sheep raisers bred ewe lambs, and quite a few still adhere to that practice. The decision is up to the individual grower. You will encounter very strong opinions on both sides of the question, but it is fair to say that the current trend is toward breeding ewes as lambs.

The change in attitude can be attributed partly to economic considerations and partly to the breed composition of modern flocks. In times gone by, profit margins were more generous than now, and few farmers took the time to sit down and calculate expenses and income with a view to optimizing profit. Such a casual attitude is not permitted in today's markets, and a ewe's lifetime production is an important factor.

The question is partly whether the lifetime production of a ewe is increased or decreased by early breeding. The argument is made that premature breeding shortens the ewe's useful life and is hard on her. I personally don't know of any controlled experiments that test that theory, but it is a popular belief. There is considerable evidence from studies by agricultural scientists in Oregon, New Zealand, and Norway that the reverse is true.

It may well be that the prejudice against breeding lambs comes down to us from experience with some of the old breeds in the north of England and in Scotland. By a combination of breed types and scarce feed, the growth of ewe lambs there is sometimes so slow that they are not bred for the first time until they are two-year-olds. The same breeds might mature much earlier on better feed, although the old Scottish "herds" (shepherds) scoff at the downs breeds as turnip-eaters who wouldn't survive on the sparse fare of the Highlands. They may be right, too, because a Suffolk ram of ours grew thin over the summer in a rich pasture he shared with Lincoln, Finn × Lincoln, and Lincoln cross rams who stayed fat and sassy.

Be that as it may, the turnip-eaters like the Suffolks do mature early, as do Finnsheep and their crosses, and if you still have rams and ewes mixed in your pens or lots of lambs, you may have already begun breeding without intending to do so.

One can argue from a biological basis that longlived species—such as elephants, tortoises, and, of course, man—tend to be slow in maturing. These slow maturers, among mammals at least, seldom produce multiple births. In contrast, rapidly maturing species like rabbits and rodents mature early, have frequent litters of offspring, and relatively short lifespans. Perhaps by selecting our flocks for rapid growth and early maturity as well as for prolificacy, we are also selecting for a short useful life. Certainly it is a fact that Finns have a shorter useful life than other major breeds, and it is a joke among Suffolk breeders that one nice thing about Suffolk rams is that people have to come back soon and buy new ones because they don't last very long.

Remember, if you do breed the ewe lambs, you probably should not tease them because teasing increases the likelihood of multiple births, and ewes lambing as yearlings may have problems with twins. Ewes bred as lambs commonly do not have enough space inside their immature bodies to hold more than one lamb by the last part of gestation, and they also will generally not produce enough milk to suckle more than one lamb adequately.

3

BREEDING

Breeding time on a sheep farm is almost as much fun as lambing time because things really are happening. The rams start pacing along fences and may get into some head-butting matches as their hormone levels climb with the season. Placed with some ewes, a good ram will rush about sniffing rear ends and curling back his upper lip after the investigation, not unlike a wine lover evaluating the nuances of odor of a fine bottle. He will lift one front leg against the side of a ewe he favors and talk to her in throaty, gargling sounds with his tongue hanging out. He will make attempts to mount his favorite, and failing that will try another and another. It is really quite a show.

EWES

Ewes who are not in heat will act irate and rush away from the ram. Those that are in heat are quite the opposite. They will face the ram, and if there is a crowd of them they all stand around him in a little group. When the ram moves they follow him everywhere. They respond to his attentions with lots of body language and stand solidly for him when he tries to mount them.

Pity the ewes who are in heat when there is no ram around. They stop eating, stand about aimlessly, and stare hopefully in the direction of a distant ram, calling out with load baas. For lack of a ram to follow, ewes in heat will commonly follow the shepherd (of either sex), a barn cat, or anything else that is handy. A ewe lamb who comes into heat without a ram around can show really strange symptoms. Some of them will display all of the vocal output of their elders, and they also run wildly around looking this way and that as the unfamiliar—and thus doubly distressing—feelings affect them. Don't panic and call your veterinarian; they'll all recover in a day or two.

Most breeds of sheep naturally come into heat or estrus in the late summer or fall of the year in north-temperate climates like those in the United States or Canada. In equatorial areas estrus is not highly seasonal. Studies of sheep many years ago in England, and more recently in France, have shown that the external factor that determines when a sheep begins sexual activity is the length of the daylight period (photoperiod). Many shepherds as well as animal scientists had concluded long ago that ewes were brought into heat by the shortening hours of daylight as fall approached. Recent research seems to suggest that the estrus cycle is set in motion by the long days of summer—specifically, the presence of light during the sixteenth hour after sunrise, and it is the long days of summer that set the timing for heat to begin in the short-day season of fall. Be that as it may, and scientists will disagree on points that seem irrelevant to the layman, all agree that length of daylight is the important control. A chemical called melatonin produced by the pineal gland changes in levels with changes in length of day, and seems to be an important factor. It is produced only during darkness, so levels increase as days get shorter. However, without prior long daylight, the increased melatonin does not seem to be effective.

There is a persistent rumor among sheep raisers that estrus is brought on by cool fall weather, and to hear folks talk, the first frosty nights of autumn bring on what appears to be an avalanche of cycling ewes. I have even read of a producer in the Midwest who thinks that ewes who spend the night in cold, low places come into heat before their cohorts who overnight on the hilltops. There is no scientific evidence whatsoever for these contentions. Sheep come into heat seasonally in regions that never get cold. Also, tens of thousands of sheep summer in the mountains of the American West. In many of those areas, the nighttime temperature drops below freezing nearly every night, yet those ewes don't come into heat. This is not to say that cool weather doesn't favor an increased fertility and libido of rams, but it has nothing to do with starting estrus in ewes.

Another factor in inducing estrus is the presence of a ram (as mentioned in the chapter on flushing). Those of you who would attempt out-of-season breeding would be well advised to use both rams and control of lighting. To date, inducing out-of-season estrus with hormones alone has had very limited success, and that at the expense of prolificacy. Choosing suitable breeds is the best bet at present.

If you are breeding ewe lambs, you probably should separate them so they can be bred last. The more growing they can do before breeding the better. It is best to keep them separate from the main flock until lambing because they should receive a more nutritious ration during gestation to allow them to keep nourishing their own bodies as well as providing for the enlarging fetus or fetuses.

RAMS

Shepherds manage breeding according to their goals and needs. Many commercial sheep raisers just put a bunch of rams in with the ewes and let nature take its course. This method is fine in some ways, but it leaves the shepherd pretty much in the dark as to the parentage of the lambs. It also eliminates any possibility of identifying the characteristics of any individual rams, such as fertility and transmission of genetic traits. I am not saying that this method is no good, but it is not my cup of tea. Since I am curious by nature, I want to know what an individual ram can do rather than what a committee can do.

A difficulty can arise with the use of a group of rams: the problem of the dominant ram. If, in a group, there is one ram that is very aggressive and is constantly trying to exercise his dominance over the others, two problems may surface. First, the dominant ram will repeatedly emphasize his superiority over his colleagues by fighting with them. This results in injury, even death, and all of the participants expend energy on fighting that should be spent on breeding ewes. The situation encourages competition among the rams—not what the sheep breeder wants at all. The breeder wants the ram's entire attention given to the ewes. Many a human female has encountered similar problems with a man who does his dominance sparring with other men in the form of bowling, softball, tennis, or some similar ritual instead of paying attention to her.

Second, the dominant ram may prevent the other rams from mounting any of the ewes in heat. Having done so, the job is left to him, and he has problems. First, having fought all of the other rams, he has expended much more effort than any one of them. He's pooped. Worse, if he does have enough energy to mount and impregnate all of the ewes he has fought for and won, then he expends a tremendous amount of semen. His sperm supply and fertility may have been fine at the outset of breeding, but because of his servicing every ewe, his sperm count begins to fall, with the result that there are fewer settled ewes, and fewer multiple fertilized ova among those that do settle.

If things are this bad, you ask, how do some breeders get away with it? They do if there is not a strongly dominant ram in the bunch. If they are all a group of fairly easy-going characters, they will spend their time with the ewes rather than trying to prove how tough they are. It helps in a ram pool to have all the rams of about the same age and breeding. The more equal they are, the less they will feel obligated to prove themselves.

In some settings, the dominant ram is so busy fighting with other rams, that some less combative rams breed ewes while the bigshots are fighting.

I don't mean to imply that rams stand around thinking all this over and make reasoned decisions as to whether to fight or not. However, a mixed group means mixed personalities. Different breeds of sheep have quite different psyches as do rams of different ages. A fun-loving young Lincoln is a far cry from a grouchy old Karakul, especially if the Karakul has a short male complex. In contrast, two of our mostly-Finn rams thought that they were the toughest kids on the block, and were promptly killed after picking fights with rams twice their size.

My preference is to put a single ram in with a group of ewes that he is able to handle comfortably. That way there is no fighting, and the sire of the lambs is known. Rams of superior performance can be identified as can those who don't quite do the job. A sterile ram will show up quickly, whereas in a ram pool you could support a sterile individual for years without even knowing about it. In the same way, rams with undesirable genetic traits can hide in a ram pool, but the bad characteristics they transmit can be traced directly to them in the sole-ram situation.

The size of a breeding group depends on a lot of factors. To test a ram lamb or to use an old, decrepit fellow you might use a handful of ewes. At the other end of the spectrum is the question of the maximum number of ewes that a single ram can settle. Various sources suggest from forty to sixty ewes, which is certainly in the right ballpark. A number like that is not the whole story, however. No one ram could possibly handle forty ewes that all came into heat on the same day. Thus, if estrus cycles have been synchronized by some method, the ewe/ram ratio needed is quite different from when estrus is spread out over many days or weeks. With synchronized ewes, the breeder either should place very small groups of ewes with each ram or should use a ram pool. If the ewes' heats are spread out, then a single ram can settle a large number of ewes. One year we used a single ram on ninety ewes and he settled every one—over a four-week period.

After a ewe has been bred once by a ram she will probably settle. This is not always the case, and most shepherds will leave ewes with the ram for at least another cycle. Another option is to use a ram to generate replacement ewes for the first cycle, then replace him with one to make growthy market lambs for the second cycle. That way your replacement ewes come from mothers who cycle early and easily.

At that point ewes can be placed with a cleanup ram to try to settle them or they can just be put with a teaser equipped with a crayon to identify the open ones. An older ewe who comes into heat after breeding should probably be shipped. You may wish to wait a little and give her a pregnancy check in a few weeks because pregnant ewes can come into heat again, although that is unusual. (I'm assuming a normal fall breeding season.) A ewe who doesn't settle in June should not be blamed.

If breeding time takes place during hot weather, the rams should be removed from the ewes during the hot part of the day and given a cool, shaded place to rest up for night breeding. This is more work, but you will be repaid by getting more out of the rams than if they were working in the hot sun.

Marking Crayons

The best way to use a ram to his maximum potential is to put ewes in with a teaser who wears a harness and marking crayon. As the ewes come into heat, they can be pulled off in small groups and put with the fertile ram. This way, the shepherd has quite a bit of control and can utilize facilities like lambing barns to their maximum by planning lambing days for groups of ewes of a size that suits the barn's capacity. The ewes can be put with the fertile ram as soon as they are marked by the teaser or they can be held over for the next heat, or even the next after that. The fertile ram should also wear a harness and crayon of a contrasting color so that the date of breeding is known. All of these details may seem like a lot of time-consuming work, but it is more than repaid at lambing when the shepherd knows the approximate birthing dates for each ewe. Also, the lambings can be spread out over a period of time without the boom-and-bust succession that typifies so many lambing seasons.

Marking crayons or blocks fit in a harness on the ram and leave a colored mark on the rump of a ewe when she is mounted and bred. Raidex, Stayfix, and Sire-Sine are three of the brands. Generally any given brand will only fit in that brand's harness. Some come in different grades for hot or cool weather. Effectiveness depends a lot on the weather and type of wool on the ewe.

For some unknown reason, the red crayons seem to mark much better in cold weather than the other colors. Black crayon marks are too easily confused with dirt or with grease rubbed off from a piece of farm machinery to be useful in all settings.

Instead of using crayons, paint or grease can be wiped onto the ram's brisket every day, but that is more of a nuisance. The crayons are not maintenance-free, however, and should be cleaned off with a pocketknife every day or two if they cease to give clear marks.

Important note: some colors of crayons will not wash out readily, so the marked part of the fleece is spoiled. Worse yet, some red crayons will tint the whole fleece when it is washed unless the marked part is removed before washing.

Fencing

Breeding is a time when your fences will be tested for soundness and effectiveness. A ram with sex on his one-track mind will view a forty-inch fence with scorn and sail over it like a deer as he responds to the siren call of a ewe in heat. Rams who are built for power rather than leaping ability will simply smash through and flatten a woven-wire fence as if it were made of wet spaghetti. We had a stubby, ornery, powerful half-Karakul ram who was separated from some ewes by a sturdy fence and a two-acre lake. One late summer day he could tolerate no more celibacy, and he just wiped out the woven wire, ran and leapt into the lake, and proceeded to swim clumsily but persistently till he finally reached the ewes. Another ram bred a ewe through a combination woven-and-barbed-wire fence without bothering to jump it. Not only did he breed and settle her, but he left a clear crayon mark on her rump. Never underestimate the inventiveness of a ram in pursuit of a ewe.

The moral is to get out and around the farm before breeding; tighten fences and keep checking them because they may get torn up. A sagging woven wire is an invitation to a ram on the move. If the fences are topped with barbed wire, be sure that the top wire is high enough to discourage jumping. A ram who leaps and misses can injure his underside pretty badly, with obviously unwanted results.

With small groups, the problem of a ram's leaving is lessened by placing the group in a small pen without enough space to give him a run at the fence. Small pens made with cattle panels are satisfactory. Whether the groups are large or small, it is never a good plan to have groups directly adjacent to one another. The rams will usually fight through the fence, messing up both themselves and the fence in the process. Ram lambs will sometimes behave in adjacent pens, but don't ever trust the older ones.

NUTRITION

During breeding the comparatively high feed levels of flushing should continue. Ewes need to be on a generous ration in order to come into estrus and in order to produce the multiple ova that will produce those valuable twins and triplets in five months. If pastures or crop residues were used for flushing, be sure to check them to see if they are still providing enough nutrition. If they have been depleted, provide supplements in the form of grain or a balanced feed mixture. A field of cornstalks with the grain all gone won't do the job, and even standing alfalfa may require some supplemental energy from grain to make a balanced ration.

If in doubt about your feeds, you can always have lab analyses made of the actual materials your sheep are eating. Feed analysis is usually available at your state's agricultural college, so ask your local extension agent. Also, many feed companies provide free analysis; inquire at your elevator or feed store. However, feed analyses may not really tell you much about the nutritional value of a given product. Chopped chicken feathers in a feed would show as protein, but they wouldn't contribute to the sheep's nutrition. The only analytical instrument that is of true value is a sheep. E. W. Crampton and L. E.

Harris in the second edition of their book *Applied Animal Nutrition* (San Francisco: W. H. Freeman, 1969) give an instructive example in which the lab analysis of a grass hay is compared with the analysis of the feces from a cow fed the same hay. The protein content of the feces is higher than that of the hay. This does not mean, of course, that we should be feeding cow pies to the stock; it means that analyses don't tell us all we need to know about the nutritional value of the material analyzed.

Perhaps the easiest route is to stick with conventional feeds like hay or hay plus grain until experience is gained with other possibilities from actual feeding trials. An experienced stockman in your area is probably the best source of information. Our pastures are pretty well shot by breeding season, so we feed about three pounds of hay and a pound of corn to our medium-sized ewes.

The flock will need fresh water, as always. If grain is fed, the sheep can get their water ration at the same time and place as their grain. Bringing sheep to the water is always easier than the other way around. This is especially true in cold weather when tanks are freezing at night. One central tank is a lot easier to keep thawed than several scattered all over the farm. Salt or a 50:50 salt and dicalcium phosphate mixture can be provided with the water. There are no special vitamin or mineral requirements during breeding, but do remember to add 1.0 to 1.5 percent limestone to the grain ration to correct the calcium and phosphorus ratio. Limestone can also be added to the salt mixture. Assume that the sheep will eat about one-third of an ounce of salt mixture daily; add limestone accordingly. I want to emphasize the need for the calcium from the limestone, because rams are in with the ewes and they are the ones who are susceptible to water belly, aka urinary calculi.

When breeding season is over—a decision that is made by the shepherd rather than the sheep—the feed levels should be reduced. During early gestation the nutritional needs of a ewe are little more than during the rebuilding period. Do not abruptly withdraw the grain or other highly nutritious components of the feed. While taking away grain is not as touchy as adding it to a ration, the rumen flora still have to adjust to a new melange of materials. This change in the proportions of various microorganisms takes a little time. The ewes will complain loudly when you reduce their grain, of course, but turn a deaf ear to their entreaties. Some authorities state that too rapid a reduction in feed levels after breeding results in poor implantation of the fertilized ova on the wall of the uterus, resulting in open ewes or fewer twins and triplets, so be sure to change proportions slowly.

During breeding season the rams are the busiest individuals, and as a result they burn more calories than the ewes. A really good ram, constantly checking the ewes, is at a very high level of activity. For this reason be sure the ram gets enough food. If a given ram is no gentleman (gentleram?) and bullies competing ewes away with his head when eating grain, you can probably rest assured that he is getting his share. If he's a perfect gentleram and eats just his share, you probably should hand-feed him some extra grain in a bucket. Remember that the ram weighs nearly twice what a ewe does and would need more nutrition for that reason alone, not to mention his extra energy expenditures. Look at the NRC tables to see what he requires, then give him 25 percent more to allow for his extra work load. You will find that some rams get so involved with romancing the ewes that they will not want to eat. This is fine, provided they came into breeding with enough flesh to be able to live partly off their own tissues. If a ram on this fasting regimen begins to look too thin, remove him from the ewes for a few hours a day so he can get his attention away from sex and onto food. Rams are basically hedonists, and it is just a matter of directing their pleasure-seeking in the right direction. An advantage in giving the ram a little extra food each day is that it presents an opportunity to examine his gear and adjust a harness or clean a crayon without having to chase him all over a pasture or yard.

MANAGEMENT OF BREEDING

Synchronization

If synchronization of heat (estrus) is part of your plans, now is the time, provided it was not done during teasing. The most common way to synchronize a flock is to use the hormone progesterone or a compound that mimics progesterone (progestogens). Progesterone is a natural hormone that prevents cycling in a pregnant ewe. The usual way is to place a hormone preparation on a plastic sponge that is inserted into the vagina of the ewe. The hormone is absorbed slowly by the ewe and prevents estrus. Alternately an implant of Syncromate can be placed in the ewe's ear. When the shepherd wants the ewe to cycle, the sponge tampons are removed by a string attached to them, or the syncromate is removed through a small cut, and the ewes come into heat in about forty-eight hours.

For the sponge method, the progestogen solution is soaked into the sponge, and the sponge is allowed to dry for a couple of days before use. The sponges can be the cylindrical ones used for some kinds of home permanents, although a veterinarian friend of mine told me he made some dandies by sharpening one end of a piece of one-inch electrical conduit and used it to cut cores out of two-inch-thick polyfoam cushion material. A piece of fishing leader is threaded through the center of the cylinder and left hanging out of the vulva after insertion into the ewe's vagina. Syncromate is inserted with a special tool.

When the sponge or implant is removed, the ewe is given an injection of FSH (follicle stimulating hormone) or PMSG (pregnant mare serum gonadotropin) to bring on estrus.

There is also a product on the market that is a chemical analog of another type of hormone called a prostaglandin, sold under the name Lutylase and made for use with cattle. Some experimental use of this product with sheep in Ireland has not been encouraging, but researchers at Agriculture Canada's Animal Research Institute report results comparable to progestogen-impregnated sponges. Trials of prostaglandin PGF2a at the Colby, Kansas, Agriculture Experiment Station by Frank Schwulst showed that it was effective for synchronizing already cycling ewes, especially older ones. Lutylase is given to cycling ewes as a single injection, or in some cases two injections, and they are immediately placed with the rams. This is a great deal handier than the sponge technique, but the cost of the hormone is presently much higher than that of the progestogens used with the sponges.

Note: None of the above synchronization methods is approved for sheep. However, you can work with a veterinarian and use the materials off label with the vet's supervision.

Progestogen-impregnated sponges are also used to make ewes superovulate, that is, to produce more ova than usual. For superovulation, PMSG is injected into the muscle after removal of the sponge. The ewe is then bred as usual after a delay of about forty-eight hours. Later, fertilized ova are flushed out of the ewe's uterus or removed surgically and are implanted in other ewes to continue their development. This procedure is strictly for professionals at present and is, as you might suppose, expensive. Transplanting of ova is widely used with cattle, especially dairy cows, to generate high-producing animals in larger quantity, faster than can be done with ordinary breeding. Because of the high cost, applications in sheep breeding are not yet common except in special instances with very valuable purebreds. In Australia, ova transplanting has been used to increase numbers of rare breeds threatened with extinction. Doubtless the technique will become more popular as costs diminish.

Artificial Insemination

Artificial insemination (AI) is used with sheep, but not as extensively as with cattle. Ram semen does not seem to store as well as that of bulls and also cannot be placed through the cervix of a ewe as easily as this job is done with the much larger cow. AI is used with sheep, but is a job for professionals. Dr. Martin Dally, University of California, Davis, mrdally@ucdavis.edu, is widely respected. Elite Genetics (see appendix 6) is a well-known firm that does AI in sheep, or ask your vet for recommendations. AI is often used with imported semen or semen collected from rare-breed rams or rams with unusually desirable characteristics.

Fertilized ova or embryos are also imported for transplantation into ewes, avoiding the need to ship whole sheep from far-off places. Also, embryo import may be allowed, whereas imported sheep would be forbidden or subject to long quarantines. Be warned that importing anything (sheep, embryos, or semen) from the United Kingdom or Europe is very tricky because of disease questions.

If you want to import genetics, do consider a whole ram as an option instead of semen. Or, import embryos, and then you will have several purebred sheep to breed naturally. Importing semen, embryos, or sheep is expensive, so compare all the costs over several years before you decide on one approach.

Year-Round Breeding

One intensive management scheme is to try to breed ewes on a year-round basis. This means selecting out the open ewes and running them through breeding again until they do settle. Most breeders do not literally want to breed and lamb continuously, but still might want to have two, three, or more lambing times per year. A rational way to start such a plan is to place ewes with a teaser permanently, and pull off breeding groups as they come into heat, starting with the least likely time of year. Why the least likely? Because you want to identify the ewes who will come into heat outside the normal season and breed them then to take advantage of their unusual trait.

Ewes can be returned to the pool as soon as their lambs are weaned, and thus produce more than if they were given a recuperation time. Breeding groups can be taken out every four months, or every three months, or every month, as suits your plan. With lambing taking place more frequently, facilities are more fully utilized, with resulting economies, at least on paper.

Such a scheme should probably include synchronization of heats so that lambings can be completed in compact time periods. Lambing can be chemically induced to space birthings even closer together; this will be discussed in chapter 5. In a French experiment, inducing was used to cause all lambs to arrive during the week so employees could have weekends and feast days off.

I mention such ideas just for you to think about. I do not suggest that a novice attempt to embark on a year-round lambing plan at the outset. Such a management scheme is strictly for the experienced, so learn first before you jump into more action than you are prepared for. Also, you may find that labor costs outweigh the gain in total lambs for the year, because you have two or three lambings to deal with. Think before you leap.

If such a program sounds interesting, check with Cornell University's Animal Science Department about the STAR system of sheep breeding: www.ansci.cornell.edu/ and search for star system or write to Department of Animal Science, Cornell University, Ithaca, N.Y. 14853-2801.

Inbreeding

Before breeding groups are assembled and put with a ram, be sure to make a last-minute check on the ancestry of all the ewes to assure they are not being bred by their sire or by a sibling. It is easy to let this happen accidentally. In the normal course of things, close inbreeding is not a good practice. Lambs from such matings are usually of low vigor and may exhibit recessive traits that were not expressed in either dam or sire. Once in a while such traits are just what the breeder is looking for, but mostly they are traits like dwarfism, blindness, aberrant wool, and even bizarre features like single eyes, no lower jaw, ears that leak milk when the lamb sucks, and so forth.

There are instances when the breeder wishes to inbreed closely to emphasize a trait. When inbreeding is intentional it is dubbed line breeding. A good example of this is the breeding of black (and colored) sheep to try to emphasize the black wool trait. The black wool gene is a recessive one for most sheep breeds, with the notable exception of the Karakul, which does carry a dominant black gene. Thus, a breeder may wish to breed a black sire to black daughters in order to bring out the recessive black trait. The genetics of wool color is very complex and far beyond the scope of this book. The Natural Colored Wool Growers Association is a good place to start to learn more. See appendix 6. (In addition, you might want to go to groups.yahoo.com and search for sheep-color-genetics, and subscribe to that mailing list.)

Such a breeding program is fine except that other recessive traits may be emphasized that are not desirable. A result of this line breeding of sheep for color is that a disproportionate number of black sheep are small or even dwarfy compared to their white counterparts.

We discovered a serious negative trait in a line of black sheep we used to have. They all carried both a recessive and dominant black gene, which was just great. However, they also showed a tendency to short legs and a propensity to die of congestive heart failure at an early age. We got out the records and traced the ancestry of the short-legged ones to a single sire. The ones that hadn't taken care of the problem themselves by dying of heart attacks were given the ultimate cure—a trip to market as culls.

Another example of close inbreeding that had a more positive outcome was carried out by Professor Leroy Boyd of Mississippi State University. He tried to develop sheep that are well adapted to the hot and humid conditions of the deep South. He had noted that animals with greater development in the loin and rump, thicker skin, and deep rather than wide bodies tended to have a greater heat tolerance. Starting with a Dorset flock, Dr. Boyd selected and bred back close relatives. His present flock at the Mississippi Agriculture and Forestry Experiment Station includes a large number of what he calls adapted animals. The entire adapted flock is traceable to four ewes and a single sire. The lambing rate of the adapted Dorset ewes is 2.5 lambs/ewe/year, almost twice that of the nonadapted Dorsets. The adapted flock is also more resistant to worms as well as being free from heat-associated sterility and abortion problems.

This "miracle" flock didn't come easily; many lambs were born defective or died from various causes, and many ewes and rams were shipped because they did not show superior characteristics. The result was a unique flock. Boyd has also established adapted flocks of Suffolks and Hampshires by the same careful selection and ruthless culling procedure. Rams from the three flocks have been sold to breeders in many states, and Boyd states that "we have not been requested to refund the purchase price or exchange animals because of reproductive failure. Many ask what we did to make them so active and fertile."

A breeder attempting to develop a specialized flock should realize it is a long and frustrating task that requires keeping only a few and losing many lambs before they ever reach market weight and return any income. The process requires money and nerve to do it right, and it is not for everyone by any means. We can all be grateful that there are a few people like Leroy Boyd who are brave enough and have the resources to try it.

Handling

Medical Care

Both ewes and rams should have been examined prior to breeding and treated for any disease. Breeding is not an appropriate time to handle the sheep much. Treatments for internal or external parasites should be avoided because the potent chemicals used may have adverse effects on the sheep's reproductive capacity. If some of the animals are parasitized during breeding, that's bad, but treating them at this time may be worse.

Similarly, vaccinations should have been done prior to breeding, although boosters at breeding are commonly given. If you have not already done so, you might consider vaccinating against a virus called parainfluenza-3 (PI-3) to aid in the reduction of respiratory diseases. The idea is to protect the animal against a viral infection in the upper respiratory system that sometimes precedes invasion by bacteria. There are both nasal and injectable vaccines against PI-3, most of which are combined with vaccines against other viruses that mostly affect cattle. These vaccines are not approved for sheep, so consult with your veterinarian before using them. The injectable vaccine causes a general viremia that can trigger abortion, so it should not be used with pregnant animals. The nasal vaccine can be used any time.

Removing Rams

The rams will have been handled a bit during breeding, but the big day is removing them from the ewes and putting them back together. They will still be excited from breeding and will take turns trying to mount one another or fight. This hassle will settle down with time, but in the meantime you don't want one or more of them injured or killed. The best solution is to confine them closely so they cannot get enough of a run at one another to do much damage. One clever suggestion by Gary Jones of Peabody, Kansas, is to put all the rams in a pickup truck with a stock rack and leave them together for a day or so. One can also pen them closely and put a bunch of old tires lying flat in the pen to trip up any ram who tries to get a run at another. There are also masks that can be put on the rams so they cannot see forward, and so can't butt. These are very effective in preventing fighting. However, when they are removed, even months later, the rams may fight immediately, so caution is advised.

Feeding

As always, be on the lookout for sheep who are off their feed. Being off feed during breeding is probably nothing serious, though. Some rams will devote themselves to their work even to the exclusion of eating. Ewes will frequently not eat just before or during their heat cycle. Don't relax altogether, however, because an isolated ewe, especially an older one, may bully her way into some extra grain and get acidosis. Treat such cases with sodium bicarbonate drenches as described in the chapter on flushing.

BREEDING AND LAMBING DATES

Breeding is the time for the shepherd who uses marking crayons to be out with notebook and pencil every day to see which ewes have been marked. It takes only a few minutes while the ewes are eating. It helps to note which ewes are standing with the ram even though they are not yet marked. If a ewe is seen consorting with the ram one day and is not marked the next, you'd better catch the ram and check his crayon.

The lambing dates will be about 146 days from the marking dates. Finn crosses have a slightly shorter gestation, and longwool breeds a slightly longer one. Knowing the approximate lambing time for each ewe is very convenient. You may find that when lambing comes there will be breaks in the sequence to give you a day or two off to sleep late or go to town. The convenience factor is not to be overlooked, because a day off during lambing can be a mighty welcome time.

In addition, there are times when knowing the lambing date can be very important. There are health conditions of the ewe that may require removing the lamb either by Cesarean section or by inducing lambing with hormones. If the ewe is known to be near term, a simple injection or two may allow you to save both the lamb and the ewe. If the ewe is not yet ready to lamb on her own, then hormones won't work, and the decision has to be made whether to remove the lamb surgically. The point is that knowing the lambing date allows the shepherd and the veterinarian to make a decision based on solid information rather than wild guesses. When you are tempted not to use marking crayons or other methods to identify breeding dates, think twice because you might regret it later if you don't take the time now.

That said, you may find that as you gain more experience, the use of crayons becomes unnecessary. As your eyes get tuned to the signs of impending lambing, you will probably give up the nuisance of crayons and the like.

THE LAMBS

Replacement ewe lambs should be taken away from the rest of the lambs so they can eat more roughage and be kept away from eager ram lambs. At this point many lambs will be ready for market, and those of you who have had lambs on a highly nutritious ration will have already shipped quite a few to start bringing in some income. Lambs raised on pasture will just now be starting to be ready. This is a time to start checking lambs for finish (see appendix 2) and for shipping when ready. Once a lamb is up to market weight and finished to grade Choice or Prime it should be moved out. It will not convert feed very efficiently after it has reached market size, so your profit margin begins to slide if you keep lambs around. Also, be sure you don't let lambs get overweight or overfinished for your local market, because you might get a lower price for them.

Many shepherds will bring ewes off pasture in the fall with lambs at their sides. Those lambs will be separated from the ewes and graded for finish. In exceptional years, range lambs may be finished to Choice condition right off the grass but usually not. When I say a range operation, I don't necessarily mean only a flock of 20,000 ewes that summers in the high mountains of the Rockies. A flock of 25 ewes that spend the summer with their lambs in Vermont or Arkansas can also fall into this category. The distinguishing feature of a range operation is turning out the ewes to fend for themselves between lambing and the next breeding season. The lambs tag along and are weaned by the ewes as they start to eat grass.

When the lambs are separated out they will either be shipped as feeders, or put onto feed by the owner to get them to market condition. If you are going to feed out your own lambs, be sure to vaccinate them against enterotoxemia using the *Clostridium perfringens* vaccine. At their age, the type D is probably sufficient. If the lambs were vaccinated previously, a single booster is sufficient. If not, they should receive a second vaccination in about two to four weeks. The vaccine for both types C and D doesn't cost much more than the plain type D if you want to be cautious.

The lambs should doubtless be wormed also before being placed on full feed. Take a fecal sample to your vet to be sure, but lambs that have been with ewes for a few months will almost certainly be carrying a worm load. You'll never convert four pounds or less of feed into a pound of lamb if you are trying to support a few thousand worms too.

Apart from worminess, and possibly infestation with lice and keds, lambs coming off the range will probably be in far better general health than if they had stayed around the home place for the summer because the opportunity for disease transmission is so much lower in the wide open spaces. Just remember to build up their feed gradually and keep an eye on them for the first few days. Then watch 'em grow.

4

EARLY GESTATION

After the highly structured and somewhat labor-intensive time of breeding, the early gestation period is very peaceful indeed. The ewes will have altogether lost interest in rams, and the rams themselves will be thin, weary, and not a little bedraggled looking. After some antediluvian combat in a pickup truck or small pen, the warrior males will go back to eating and sleeping for the rest of the year, the social structure of the world of rams having once more been established.

For their part, the ewes will be united into one or a couple of groups again and will have a wonderful time seeing old friends and doing some ewe-style head butting to reaffirm old pecking orders.

The shepherd will have to face the problem of cold weather a bit more seriously in the northern states, and the first snow or rain may signal the approach of winter. The shepherd and his family may become concerned with firewood supplies, Christmas presents that have to be made, the last bit of garden produce that needs canning or freezing, and a million other things that command attention, but the sheep are heading into their favorite time of year.

What more could a sheep ask for? The hot weather is finally out of the way, and they are at last comfortable in their wool wraparounds. The flies that have been after them all summer long have vanished as quickly as they came, along with those pesky mosquitoes that sought out the bare spots. There is no longer thunder and lightning to terrify the flock into huddling together. To be sure, there is no more tasty pasture either, but that is more than made up for by the fragrant bales of hay the shepherd opens to feed to them. Fresh, moist pasture is great, but there is really something special about hay. There is nothing quite like exploring with lips and tongue for the softest and most tender alfalfa leaves before getting down to the serious business of eating the stems. Yes, this is a pretty good time to be a sheep.

NUTRITION

"Living Off Their Backs"

The ewes have been on a highly nutritious ration since the start of flushing and should have come out of breeding in good condition. The high feed level during the early part of pregnancy was to maintain pregnancy, but once the ova are firmly implanted in the uterine wall, the ewes can be put on almost a maintenance diet while cell division is starting to make a lamb out of that speck of life called a fertilized ovum. After fertilization the ovum undergoes cell division, and moves from the oviduct to the uterus at about day 4–6. By day 11 the embryo is about 1 to 1.5 mm in size. At about day 12–13 the embryo becomes attached to the uterine wall. From about days 11 to 34 is the period when major organs and tissue develop from the beginning mass of cells. That is a time when one must be especially careful about medications or poisonous plants that can cause deformities. For example, the wormer albendazole administered at this time can cause fetal deformity. From that time on, the fetus just very slowly grows larger and there is less danger of deformities developing because of external causes.

For the first three months the ewes will live off a ration that will require them to use some of their stored resources. This so-called living off their backs means that they will lose some weight during early gestation even though the fetus is growing inside their wombs. The purpose of limiting feed at this stage is to ensure that the ewes do not go into the last four to six weeks of gestation in a fat condition. Female readers may recall being given the same advice during a pregnancy of their own.

Ewes can be put on a diet at this time because most of the growth in the size of the fetus does not take place in the early stages of gestation. The fetus grows in complexity as cells become differentiated, but a large increase in weight of tissue does not accompany the specialization of the cells. One might say that there is a big qualitative change in the fetal tissues but not much quantitative change.

As a result, the flock will get by just fine with about three to four pounds of average-quality hay per head per day. If hay is not fed, a suitable ration can be calculated from the NRC tables. This should provide adequate but not excessive total digestible nutrients and the right balance of energy and protein. This is the time to begin to serve up your hay in the correct order. Start with your poorest hay (which should be a little leafy and have at least 8 percent protein) and gradually change over to better and better quality as gestation progresses, saving the better stuff for late gestation and the very best, the highest protein hay, for lactation.

Bred lambs should not be put on the weight-loss diet. You want them to gain some weight in early gestation. Many of them wouldn't consume three to four pounds of hay anyway without wasting a lot of it, so feed them about two pounds of hay and two pounds of corn to keep them growing.

Provide all of the ewes with a mineral mix or salt plus dical (dicalcium phosphate) and keep the water available.

Depending on prices in your region, you might want to replace part of the hay fed with corn for the older ewes too. Corn has roughly double the nutritional value of hay, and in some areas can cost about the same per pound. Also, protein needs of ewes in early gestation are low. A couple of pounds of average-quality hay and a pound of corn will handle an average-sized ewe just fine, and will save hay and possibly money as well.

Crop Residues

Some sheep owners (I won't call them shepherds) think that a ewe can get by during gestation on feed that is practically no feed at all. That is taking the idea of a ewe's living off her back a little too literally. For example, some people turn the flock into a field of cornstalks and hope for the best. What will happen in such a case is that the ewes will seek out the spilled grain first and this may result in some deaths from acidosis. The survivors from that episode will then gradually lose condition as they try to endure on a diet that contains only about 2 percent protein. A 150-pound ewe would have to eat between eight and nine pounds of cornstalks a day to get along. Even if she could manage to stuff that much into herself, which is doubtless impossible, digestion would be slow because the rumen flora need a minimum amount of nitrogen from protein to digest the stalks. The same thing can be said for straw, especially wheat straw. In either case, the roughage of stalks or straw would need to be supplemented because of a lack of energy and a big lack in protein.

If straw is considered as a feed, there are two things to evaluate. First, is the straw actually a cheap feed on a per-pound basis? Straw may be cheap by the bale, but straw bales don't weigh very much. Second, is the combination of straw plus a grain and protein supplement actually cheaper than just feeding hay instead? Get out that calculator and pencil and paper.

If straw is inexpensive, it can be used as a sort of filler to blend with more nutritious feedstuffs to give a mixture that can be fed free choice. Chopped or ground straw mixed with hay, grain, and oilseed meal gives a mixture that fills up the ewe before she can overeat. Some experiments with straw mixtures at the Hettinger Experiment Station in North Dakota, in the heart of wheat straw country, have indicated that they are a practical option to reduce the labor of feeding during gestation, although mixtures with too much straw gave poor results. A grower would have to experiment a bit to get the right blend.

Another thing that can be done with straw is to treat it with anhydrous ammonia to add some nitrogen and thus increase its feed value for sheep. Stacks of straw bales can be covered with six-mil plastic sheeting and edges sealed with soil, tires, lumber, or other weights. Ammonia is then introduced at a rate of 60 lb/ton over the course of twelve to twenty-four hours, after which the stacks are left to react with the ammonia for three to four weeks.

In a trial at the Colby, Kansas, Experiment Station, ammoniation increased crude protein of the straw from 6.5 to a final 11.2 percent, a whopping 72 percent increase. Digestibility as measured in the laboratory increased from 30 to 48 percent. Ewes fed the ammoniated straw in a forty-day trial had average daily gains of 0.28 pounds compared to 0.18 pounds for ewes fed untreated straw and 0.47 pounds for ewes fed a control diet of silage, grain, and hay. If you raise sheep in an area where straw is inexpensive, ammoniation might be a worthwhile consideration.

Stockpiling

Stockpiling means letting standing pasture mature for later grazing. In much of the United States if a pasture or hayfield can be kept ungrazed from early August, then there will be enough growth for winter grazing. Fescue, red clover, alfalfa, and trefoil are all good stockpile crops. Grazing of the stockpiled feed should be controlled by portable electric fencing to limit intake. Inquire locally for practices in your area.

Hay Feeders

If hay is a major ingredient of your feed, some consideration has to be given to how to feed it. One very important thing to keep in mind is feeding in such a way as to minimize contamination of the ewes' wool with leaf and other vegetable debris from the hay, the bane of fleece cleanliness. Feeders that allow the ewes to burrow into the hay are to be avoided at all costs as the neck wool will fill with trash, reducing the value of the fleece for any purpose and making it altogether worthless for handspinning. In a lot or confinement setup the feeders should be designed so the ewe has to eat down into them over a barrier. She also should have to stick her head under a board or something else that discourages her from lifting her head up with a big mouthful of hay and dribbling leaf into her neighbor's neck wool.

If feeding is done in pastures, consideration can be given to feeding on the ground, even though many authorities tell us that we shouldn't do that. Here in Minnesota, pastures are snow covered for up to five months a year during a time that coincides with gestation, so we feed hay on top of clean snow. While there is theoretically a hazard of disease transmission if the sheep urinate and defecate over the feed, we have experienced no such problems. Feeding on the ground does give very clean wool, which for our flock is of great interest. In other localities, or with other flocks, feeding on the ground might not be desirable.

Feeding Schedules

During early gestation when the sheep are given less hay than they would eat free choice, feeding them is a daily chore. Not that checking the flock every day is not a good idea, but it is nice to stay inside sometimes and let the critters take care of themselves. Bob Jordan and others at the University of Minnesota had the same idea. They had to hire people to feed the sheep on weekends, which was an added expense, so they tried feeding only on Monday, Wednesday, and Friday of each week. Their formal feeding trial compared the

This common type of hay feeder encourages burrowing, which the center sheep is doing. Burrowing results in vegetable-matter contamination of the neck wool.

three-day group with a control group fed the same total amount of hay, but fed every day. They found that not only did less frequent feeding make no difference, but that the group did even slightly better on average in terms of weight gain of both ewes and lambs.

We have fed our flock on an every-other-day basis for many years and are very happy with the program. We just feed double the daily ration on alternate days. The sheep quickly learn the new schedule and adapt readily. In fact, they seem to be more content than with daily feeding. The big eaters get full and have to quit, which gives the slow and shy ones a crack at their fair share. When we now and then give them a one-day ration in order to shift our schedule for some other reason, the sheep miss their double shot and complain loudly to let us know their objections in no uncertain terms. If we are feeding grain for some reason, we feed it daily even if hay is fed every other day.

With cold weather, in many parts of the country the shepherd has to get busy keeping water from freezing. Floating tank heaters can be used, or there are various types of enclosed waterers and demand waterers that can be equipped with heaters. Check local suppliers and ads in sheep magazines for choices. Any heated waterer will cost something in electricity to operate.

Another solution is to let the ewes get by on snow alone. If clean snow is available in your pastures, it can make a fine substitute for water. The sheep will use a little extra fuel to melt the snow inside their bodies, but the cost of the extra feed needed can be shown to be less than the cost of the electricity to heat water in a tank. In a formal study at the University of Alberta, A. A. Degen and B. A. Young found that lactating ewes fared equally well on water or snow, even in the frigid environment of Edmonton. If you use snow as a sole source of water, be sure there really is plenty of clean snow accessible to them, and not only dirty, icy drifts in fence corners.

I can report from personal experience that sheep prefer snow over water when given the choice. I have learned the hard way never to fill their water tank if they have fresh, clean snow to eat. It is disheartening to fill a tank at ten below and then have to bail it empty before it freezes solid because the sheep prefer to eat their natural sherbet.

Environment

The choice of where sheep are kept during early gestation is pretty much a matter of the shepherd's convenience, with some preferring to drylot them while others will leave them in conveniently located pastures. Given a choice, I think pastures are preferable because crowding is never a good idea. In a confined area, spread of disease is encouraged and the wool gets dirtied and contaminated by vegetable matter, mud, and excrement. In pastures, the sheep can move about to get exercise and also can eat or pick at dry grass that isn't buried by snow. Convenience of feeding can be enhanced by pasturing the flock close to hay storage if the layout allows this.

We are strong believers in keeping the sheep out of the barn for most of the year. They are well equipped to cope with cold weather and thrive in it. We know sheep raisers who keep their ewes confined in the winter, and most of their flocks have persistent respiratory problems as a result. In a barn their feet don't wear down very much either, unless an expanded metal floor is used, with the result that hooves need more frequent trimming or the risk of foot disease is increased. I could go on for pages describing the health prob-

.ems that arise from confinement, but I think it suffices to say that a shepherd should provide for the sheep just what a physician would prescribe for a human patient: plenty of wholesome food, fresh air, and exercise.

Confinement

Sheep in confinement are not necessarily doomed to be in a state of poor health at all times. A well-designed confinement facility can provide a healthy environment. The average farm barn is not a well-designed confinement facility, however. Sheep housing for total confinement must provide lots of fresh air at optimum humidity without buildup of pollutants such as ammonia, and it must be free from drafts. Such a goal is easier to set than to achieve, as many a farmer and agriculture experiment station has discovered.

The naturally gregarious nature of sheep makes them good candidates for confinement raising from the point of view of their psyche. However, the lack of exercise and the ease with which communicable disease can be spread makes health care a far more critical factor than with a pastured or even a dry-lotted flock. Nutrition must be provided and monitored much more skillfully than with a free-ranging flock that has a choice of a wide variety of plants and other materials to eat. Vitamin D that would be provided by sunlight to an outdoor flock has to be given to the indoor group in some other way. Trace elements that might be present in the pasture setting must be provided in the confinement barn. In addition, the shepherd is faced with the necessity of keeping a confinement facility in constant use because of the large capital investment in the fixed costs of the building and its support facilities, plus the cost of skilled personnel to operate such an artificial environment successfully.

Confinement animal raising is to an outdoor environment what a greenhouse is to a garden. Or, put another way, would you rather raise bass in a bathtub or a pond? Some people raise commercial vegetables in greenhouses, and others grow market fish in artificial rearing facilities, but only the experts make a go of it. At present, the main advantage of confinement sheep raising is that it can be used as a tax shelter. I know that there are those who argue that protection of sheep from predators justifies confinement. According to my calculations, guard dogs, electric fences, and alert shepherds are a lot cheaper.

Others argue that confinement permits raising of sheep in climates where they otherwise would do poorly. In Canada, for example, Agriculture Canada built a large, well-conceived experimental facility that is a model of its genre. The justification is partly the cold weather over most of Canada. I happen to think that cold is more a problem for shepherds than for sheep. At the other end of the spectrum, excess heat can cause troubles such as heat exhaustion, infertility, reduced resistance to disease and parasitism, and reduction of wool growth. One solution is to confine the sheep and to cool the building with air conditioning, as is done for pigs. A better solution is to breed the sheep to tolerate heat as Leroy Boyd of Mississippi State University has done. When he read a report by some agricultural engineers who stated that a new pig building needed better air distribution and cooling because 70 percent of the pigs were doing poorly in the building, Boyd commented that he personally would cull the 70 percent and keep the 30 percent to produce more pigs who could tolerate the heat.

My personal bent is to want to control every stage of sheep raising, and I find the concept of confinement very appealing. However, that is just a daydream, and I know in my heart that raising sheep outdoors is the most reasonable approach. Our flock, when given the choice, will choose the outdoors over shelter almost every time and the more open the better. In really fierce blizzards they will bed down in the lee of a building or other windbreak, and they will seek the shade of a grove of trees on a scorching day, but otherwise they cast their votes for the unconfined life.

Maintenance

Early gestation is an undemanding time for the shepherd, so it is a good time to repair and clean equipment. Gates and panels for the lambing area should be mended, disinfected, and otherwise put in order. The barn itself can be cleaned and disinfected, and bedding laid down. Inexpensive and effective disinfectants for equipment and barn walls are diluted household bleach (Clorox, Hilex, etc.), made by mixing the straight bleach with an equal amount of water, or a commercial product called DC&R that is also mixed with water before application. The bleach releases chlorine and monatomic oxygen, and the DC&R releases formaldehyde. Either is a good choice for spraying all of the equipment to get it ready for lambing. The floor of the barn can also be heavily limed with slaked lime (calcium hydroxide, hydrated lime, $Ca(OH)_2$), before bedding. The lime is a mild disinfectant, and it also encourages breakdown of the manure and bedding by keeping the pH "sweet" or alkaline, which in turn reduces odors and harmful gases in the barn atmosphere. Do not use the plain ground limestone that is sometimes sold under the name of barn lime, as it is useless as a disinfectant. If you have reason to believe that the barn harbors pathogenic organisms such as the one that causes foot rot, you might consider liming with quicklime (burned lime, CaO, unslaked lime). Quicklime is much more caustic and reactive than slaked lime and should be treated with respect. Wear gloves, a face mask, and eye protection when applying it, and bed the barn and leave it empty for at least a couple of weeks after use.

As long as you have the slaked lime bag open you might want to consider mixing up some whitewash to paint the inside of the barn. It brightens things up considerably and has a mild disinfectant effect too. Directions for mixing and using whitewash should be on the bag of slaked lime.

With the barn in order, the start of lambing won't catch you rushing around in a panic trying to clean things up and trying to locate all the equipment you need at the last minute. Also, by bedding the barn down well ahead of time, you will give the straw or other bedding material ample time to settle. This is particularly important if the barn is to be used for shearing. A freshly bedded barn results in a lot of straw pieces in the wool, and this greatly reduces the value of the fleece, even to the point of ruining it for handspinning. If the barn is to be used for shearing, let me suggest using coarse grass hay rather than straw for bedding. We use hay (locally called swamp hay) that is mostly reed canarygrass and other similar species. Dry cattails are as good or better, if available. If you don't do your own shearing, now is a good time to arrange for a shearer to come.

Handling

Moving Sheep

Moving a flock of sheep where you want them to go can be simple or it can be very difficult indeed. Corrals, pens, sorting chutes, and a border collie make the job somewhat easier, but understanding a sheep's mind is the most helpful single element. Sheep don't do a lot of reasoning or thinking, but they do have a full stock of instincts and good memories. If you can use their natural behavior to your own benefit, you will have an easy job. If you try to make them do something that doesn't fit with their instincts, you'll have a terrible time of it, and you probably won't get the job done.

Sheep have a few strong instincts that are fundamental to their unique character. They are gregarious for the most part (with the exception of many of the British breeds). They are fearful of the unfamiliar and of aggressive dogs. A familiar environment is their favorite, although a ewe with wanderlust will now and then lead the whole flock on a voyage of exploration. Identify that individual and sell it to someone you don't particularly like, or send it to market.

Sheep are followers, and the shepherd should try to establish himself as the leader whom they will follow. To begin with, call to them in some simple way and carry a bucket of grain. If they are not accustomed to a grain bucket, spill some grain on the ground so they get the idea. They will catch on quickly and follow you anywhere you want to take them—at least almost anywhere. Keep your back toward them and walk away. If you stop to look at them, they will stop to look at you. Once they get used to following you, the grain becomes unnecessary.

The flock will avoid places where they feel trapped, such as fence corners or inside buildings. Here again the use of a food bribe is most helpful. A sheep dog is worth its weight in gold when trying to make a flock go into a building, though a couple of people clapping can do a fair job. Sheep like light too, so if the barn has an open door at the far end, they will enter more readily.

The most important factor in moving sheep is practice. The more they are moved in the same way, the easier it becomes. As I type this part of the manuscript I am watching Teresa out the window as she moves about 150 lambs from a small pasture back to their drylot. They have done this every day for weeks, and they know that the feeders will be full of freshly ground feed when they get back, so all she does is clap her hands a few times and talk to them, and they all rush, leaping and cavorting to the gate and into the yard.

Once trained to follow a given shepherd, sheep won't want to be shooed away by the same person. Our flock is trained to follow me as leader, while our border collie or a family member brings up the rear to encourage stragglers. This is a fine system except for the times when I am in the truck alone and want to go through a gate that is blocked by some sheep. I have a terrible time because they are predisposed to follow me, and all my clapping and shouting has little effect because every time I turn my back to go back to the truck, they follow me.

The problem of getting through the gate is compounded by another element of sheep behavior: they love to go through gates. Never mind what is on the other side of the gate, they want to go through. A road to a gate is almost equally compelling. With enough strategically placed gates, you can do almost anything with a flock. I always find it amusing to call sheep to a gate and let them crowd together for a moment, then open the gate and watch them rush through madly. Then, having gone through, they all stop and look back as if to say with some sort of collective puzzlement, "Now why did we do that?"

General Physical Condition

During the early gestation time, individual sheep should be examined for their general physical condition at least once. This can be done when they are being handled for some other reason such as worming or immunization. Feel their backbones and hip bones. A backbone that is a half inch or so above the flesh is normal, but a more exposed one that is accompanied by knobby hip bones means that the ewe is not getting enough food. These individuals should be removed and placed on a higher feed level. Take special care with the old ewes and the best milkers for they are the ones most likely to be in thin condition.

At the opposite extreme, ewes with their backbones and hips buried in flesh and fat should be put on a more restricted diet. If facilities allow, the flock can be divided into three groups and each subflock given an appropriate diet. The thin ones and the bred lambs can be given extra feed, the normal group left alone, and the fatties put on a reducing plan. Be sure to check hooves as you examine the sheep, and trim them if needed.

Pregnancy Checking

Many producers check their ewes at some point to see if they really are pregnant. If a teaser ram with a crayon was placed with the flock after breeding, then some indication has already been given, although remember that all this tells you is that the ewe did not come into standing heat again, not that she is pregnant. She may not come into heat because of pregnancy, but she may have also quit cycling for some other reason. Conversely, a ewe who shows heat after breeding may actually be pregnant, although this is not usually so.

It is a good management practice to check the flock for pregnancy during early gestation. There are a number of reasons to do this, but they all boil down to saving money. If a ewe is not pregnant, she is not performing her primary function. One option is to try to rebreed the open ewes, as one would do with year-round lambing. Another, more common one, is to ship her. An open lamb might be kept and given a second chance at another breeding season, but the ones who have bred before and are open now should be disposed of. Any open ewes that are retained should be kept separate from the pregnant ewes because they require less food and are too playful to be kept with ewes who are in the later stages of pregnancy.

There are a variety of techniques that can be used to check ewes for pregnancy. This was not the case a few years ago, and pregnancy evaluation was limited to larger species such as horses and cows. With such animals, a veterinarian could insert a gloved hand into the rectum and palpate (feel) the uterus to determine whether or not a fetus was present. A ewe is too small for this technique.

The first commonly used technique for sheep was to tie the ewe in a sheep "chair" and make an incision in the abdomen that allowed the veterinarian to insert his fingers and palpate the uterus. The ewe was then stitched up and turned out with the flock. It is a testament to the resilience of a healthy ewe that complications from this major surgery were rare.

Later, a less drastic technique was developed that included use of a probe that was inserted up the rectum and used to nudge the uterus against the abdominal wall so that it could be palpated from outside the belly. This approach, sometimes called the broomstick method because of the use of a broom handle as a probe by some rough-and-ready practitioners, is not without problems. The main problem is the chance that the wall of the rectum can be punctured, resulting in peritonitis and almost certain death of the sheep.

Electronics came to the rescue, and there are now simple and effective instruments to check pregnancy. The first electronic method used a probe that emitted an ultrasound signal that reflected from the fetal heart. The operator could then detect the heartbeats of the lamb or lambs as separate sounds from the ewe's heartbeat. This gadget had the disadvantage of high cost and risk of perforation of the rectum, as with the rectal probe.

Current instruments use a probe that is placed against the wool-free skin under the right rear leg and aimed up and forward at the uterus. The high frequency sound then travels through the tissue and reflects from the fluid-filled uterus of a pregnant ewe. The echo is detected by the probe, and the instrument gives a signal to indicate pregnancy. If no echo is received, the instrument signals an open animal. These gadgets work after about the sixtieth day

*A helper holds the hind leg of a
ewe as the shepherd uses a
handheld pregnancy checker.*

of pregnancy, and are about 85 to 95 percent accurate. The highest accuracy is achieved if the open ewes are rechecked. We have found that our instrument is better than 95 percent accurate with our flock. We did not achieve such accurate results with another flock that was run through rather hurriedly.

The electronic pregnancy detectors have steadily dropped in cost since their introduction and can now be had for less than $300. This may seem like a lot of money, but consider their value carefully. Also consider sharing the cost of the instruments with another shepherd or two. The money that can be saved by purchasing a pregnancy detector is substantial. A ewe checked at sixty days and found to be open has about eighty days left before lambing time when it would be discovered that she was not pregnant. During that time she'll eat about 250 pounds of hay and about 70 pounds of grain. With hay at about $60 a ton and grain at $0.05 a pound, that's $11 worth of feed that is wasted, not to mention the medications and time spent on her care. Even just finding a few open ewes a year, shipping them and saving their feed costs pays for a pregnancy detector pretty quickly. I'll refer to the economics of pregnancy detection in ewe lambs a bit further on, under Evaluation.

Better yet, use ultrasound to actually see how many fetuses are in the ewes. This allows one to feed those with singles less than average, and those with three or more, more than average. It is also preferable to have ewes separated into groups before lambing to minimize doing lots of sorting after lambing, which disturbs new lambs. The average producer will not want to buy the equipment to do this, because it costs many thousands of dollars and requires considerable skill to use effectively. There are persons who are in the business of doing ultrasound scanning and who will come to your farm to do the work. The charges can be high, but you will probably come out ahead by feeding correctly based on solid knowledge. Give it serious thought, and base your feed program on the results. Ask your vet or other producers for names of technicians who do scanning.

MEDICAL

Abortion

The threat of abortion is a constant worry for the shepherd. It should be realized that abortion is a symptom, not a disease, and every effort should be made to avoid the conditions which can bring it about.

One cause is mechanical damage to the ewe, although this is not a major problem in early gestation. Crowding through gates or doors is a major cause, as is roughhousing among the ewes. Open ewes are a lot more mobile and may be overly playful and aggressive with their pregnant colleagues. They should be kept apart.

Feeding of moldy feedstuffs is another cause of abortion. Some molds are harmless, but others produce toxins (poisons) called mycotoxins that can stimulate abortion. In an ideal world, the shepherd would never feed moldy hay, but of course in real life there is usually some mold in all but the most perfect hay, at least in areas where the rainy season and the haying season coincide. As any farmer in such areas knows, the advice to make hay while the sun shines is a wonderful idea that he does not always achieve in spite of avid listening to weather forecasts and careful planning. Each producer will have to decide what to do with less-than-perfect hay. I freely confess to having fed some moldy hay each year as a matter of expedience, but not without some nervousness. The practice is not recommended.

Abortions are also caused idiopathically, as any veterinarian can tell you. I hope the vet will tell you this with a twinkle in his or her eye because it means that the cause is unknown or, literally, that the condition is generated by the ewe herself and not attributable to an external cause. We have had an average of one abortion a year. We always send the aborted fetus and some associated tissues and a blood sample if we have one to a veterinary diagnostic laboratory. The lab has never yet determined a cause, which our veterinarian tells us is par for the course. This does not mean that the lab personnel are incompetent, it just means that many abortions are of undetectable, and therefore unknown, origin. In spite of this dismal track record, I still think the lab examination is well worth the effort and expense, because a single abortion could be the first of a major outbreak that might be stopped by early detection.

Worms

The life cycle of the common sheep worms was outlined in the Building and Rebuilding chapter. An important part of the cycle is the deposition of oocysts in the sheep's feces on the ground. With warmth and moisture, larvae hatch and grow and reinfect the sheep when eaten with grazed forage. If gestation coincides with a cold or dry time of year, the oocysts may not hatch, or if they do the larvae soon perish. In addition, if the flock is eating just hay and grain, the risk of reinfestation is greatly reduced. Either way, the reproductive cycle of the worms is effectively broken.

Many people survive the rigors of winter by moving temporarily to a warmer region. Sheep worms don't have this option, so they winter over in a dormant or arrested state in the sheep's gut. There is no reason that a sheep should be expected to provide winter housing to these scoundrels, and parasitologist Dr. Rupert Herd of Ohio State University has found that a single worming with levamisole kills 98 to 100 percent of the susceptible dormant worms. If reinfestation can be avoided, the sheep will remain worm free until they eat larvae in a pasture the following spring or summer. The worms will be in this dormant stage for flocks in early gestation that were bred in late summer or early fall in cold climates. Worming can be combined with vaccinations for vibrio, enterotoxemia, or other diseases. If levamisole is used, combining the worming with vaccination has an added advantage because levamisole is a so-called immunopotentiator, which means it causes the animal's immune system to react more strongly and faster to the vaccination. Thus, the effectiveness of the vaccine is enhanced.

If reinfestation prior to lambing is a possibility, a second worming just before lambing is recommended (as will be discussed in the next chapter).

In climates where winter dormancy is not part of the life cycle of the worms, worming is not so simple. In climates such as that of the southeastern United States, worming has traditionally been done very frequently, as often as every two weeks. At that rate, the cost of wormer alone approaches ten dollars per ewe per year, an expense that cannot be tolerated if profit is one of the motives for raising sheep. Such frequent worming also encourages development of resistant worms, so is a losing strategy in the long run.

Professor Leroy Boyd of Mississippi State University has found that his flocks that are adapted to local conditions need to be wormed only five or six times a year. Descendants of the original sheep brought to the southeast by the Spaniards are essentially totally worm resistant, and a flock at the University of Florida has not been wormed for over thirty years. Similarly, the Barbados Blackbelly sheep is highly worm resistant. Selection for worm resistance can make sheep raising a practical enterprise in almost any climate, even though the shepherd in the northern United States and Canada does have the advantage of being able to catch the worms napping, as it were, during their winter dormancy.

Vaccinations

Early gestation is the time to give the second vaccination for vibrio to ewes who were vaccinated for the first time at breeding time. If you give an annual booster, you should also have vaccinated at breeding. A shot can still be given in early gestation for some protection. Some producers who have never had trouble with vibrio wait to vaccinate until an outbreak occurs. This is really far too late because immunity takes a couple of weeks to develop, and many ewes already will have been infected when the first abortions happen.

Vaccination against vibriosis is not mandatory because the disease is not present in all regions. A closed flock that is not located near other sheep flocks may be safe even in regions where the disease is endemic. The decision whether to vaccinate against a given disease is ultimately that of the producer who must make the judgment based on discussions with a veterinarian and area sheep producers. It is a matter of cost effectiveness. Is the cost of the vaccination greater or less than the possible economic loss from the disease? That is the question you'll have to answer. One cannot vaccinate against every disease.

Early gestation is also the time to vaccinate unprotected ewes against *Clostridium perfringens* types C and D. If replacement ewe lambs were vaccinated one or more times during their first six months of life, no additional vaccination is needed at present. Purchased ewes of unknown medical history should be vaccinated now to build their immunity for maximum response to another vaccination in late gestation. In my opinion, vaccination against *Cl. perfringens* is not a matter of choice or judgment but a matter of necessity. This bacterium probably causes more deaths of lambs and yearlings than all other diseases, with the possible exception of pneumonia. A single dose of vaccine costs about seven cents, so do use it.

If tetanus is a problem on your farm, the flock can be vaccinated against *Cl. tetani*. The vaccine is fairly expensive, so don't do it if it is not called for. If you have experienced losses, you may have little choice but to vaccinate. The spores of *Cl. tetani* are present in most soil, although rarely in the northern Rocky Mountains, for some unknown reason. If you have horses on the farm, horse manure is especially rich in spores, so keep sheep away from the horses or vaccinate.

OBSERVATIONS

During early gestation the flock doesn't require a great deal of care, but be alert for any signals of illness such as a ewe going off feed. Almost any illness can be the cause of abortion or death and resorption of the fetus. Pull out any sick ewes and treat them in a sick-bay area for a few days until they brighten up.

Be alert for any vaginal discharges that might indicate trouble in the reproductive system. Walk quickly along behind the ewes while they are eating to give the rear ends a quick appraisal. Also, be sure to be on the lookout for aborted fetuses. This is easier said than done because the fetus is often covered with dirt or snow or tucked against a fence. The ewe who aborted is probably a better clue. She will act very upset, running about and baa-ing frequently. She will usually have some placenta hanging out of her vulva or at least may have wet or dark stains around the vulva area. If you find such a ewe, isolate her immediately and try to locate the fetus and any associated tissues or fluids and dispose of them by burial or burning. The curious ewes will sniff or lick at the aborted fetus and tissues or at the aborted ewe's rear, and may become infected if there are any abortion-producing organisms present. Remember that the aborted fetus and tissues can be sent to a diagnostic lab if you wish to do so.

Be on the lookout for limping sheep who might have foot rot. It is more common in fall and winter when sheep are confined closely in wet, muddy lots.

If silage is fed to your flock, you should be alert for a disease called listeriosis. Listeriosis is sometimes called circling disease because affected sheep often walk around in small circles with their heads turned to one side. The ill animals also get a fever, sit or lie down, and may drool out of one corner of the mouth because of paralysis of facial muscles on that side. This condition is caused by a neurotoxin (nerve poison) produced by a bacterium called *Listeria monocytogenes* that thrives in fermenting plant matter. It grows in silage that is improperly packed so that air can reach the growing bacteria. Properly made silage ferments anaerobically, that is, in the absence of oxygen. *L. monocytogenes* can grow in any aerobically fermented plant material, and we experienced the disease in a sheep that ate old, mowed grass in a pasture, although this is not a common source.

The only treatment for listeriosis is sulfas or broad-spectrum antibiotics, but the treatment is usually futile because by the time symptoms appear the disease has already done irreparable damage to the brain and nervous system. One problem with treatment is that it is very difficult to get high levels of an antibiotic in the brain because the sheep has protective mechanisms that inhibit transfer of medications or other chemicals from the body to the brain. Prevention by not feeding spoiled silage and keeping affected animals from contaminating feed and water is the best approach. Low levels of antibiotics in feed are used by some farmers. Exercise care in handling affected animals and tissues because listeriosis can affect humans. If you suspect that listeriosis might have been the cause of death of a sheep, leave the necropsy to the lab rather than risking infecting yourself.

EVALUATION

Rams

With breeding over and pregnancy checks made, decide which if any rams should be culled. A ram who settled only a few ewes or none at all should be shipped right now. If a ram is not fully sterile, but just of low fertility because of age or general health, he should be shipped unless he has unusually valuable genes that justify his staying on even at a reduced performance level.

Ewes

A ewe who didn't settle is a prime candidate for the truck. Even if she is one of your old favorites she probably should go unless you can afford to keep nonproductive animals in your flock.

Ewe Lambs

Open lambs present a different case. Many growers do not believe in breeding lambs, which is one way of avoiding the issue altogether. If, however, you do breed ewes as lambs, and they show open at a pregnancy check or are repeatedly marked by a cleanup or teaser ram, then what do you do?

The simplest thing is to keep the lambs over until next year and try again. This is a bit of a nuisance because they will have to be kept separate from the bred ewes so their youthful boisterousness doesn't cause mechanical abortions in the pregnant ewes. In this situation you could have as many as four different groups of ewes if you have already sorted as to feed requirements.

Another possibility is to keep the open ewe lambs with a ram until they do settle. That extends the lambing period way beyond the normal time and may mean that you will have lambs arriving when you are busy with seasonal activities such as haying or tending crops. If, on the other hand, you are trying to spread lambing out, it may be the best solution. The fact that a ewe bred for the first time in February doesn't mean that she will ever do it again. Unless your open ewe has a lot of Dorset, Merino, or Rambouillet blood in her, she will probably not recycle in late winter as her next breeding season but will hold over to the following fall, so no long-term gain is made by the late breeding. If her first lamb was weaned very early, you might be able to get the ewe bred in her second fall.

The third possibility is to cull the lamb from the flock. As I mentioned in the chapter on flushing, culling the lambs that don't settle means selecting replacement ewes on the basis of early maturity. If this is what you want, then

Some hungry ewe lambs follow a wheel rut to a meal of hay served on a pristine bed of snow.

that's fine, but be aware of what you are doing. If you already selected the larger ewe lambs to breed, then you have also selected for rapid growth. Some breeders also select their replacements only from a pool of twins, triplets, and quads. This adds still another factor you are selecting for by your choice of re-placements. It may be that you really should be selecting for wool type, or meat conformation, or resistance to hot weather, or any one of hundreds of heritable characteristics. If you selected a ewe lamb for some important char-acteristics before she was put with the ram, you may want to keep her in any case rather than cull her on the single issue of early sexual maturity.

For purely short-term economic reasons you should ship open ewe lambs. (Notice that this means ignoring the other factors if that is your choice.) First of all, you will save feed money. A lamb should be kept on a growing diet her first year, so she probably would get some extra feed such as grain during early gestation. If she was given a pound a day of grain at five cents a pound, she would eat about five dollars more feed than an older ewe who would eat about eleven dollars worth of feed in the same period. By lambing time you would have put a total of sixteen dollars of feed into each open ewe lamb, and yet get no lamb from her to justify the feed cost.

You should also consider the selling price of the lamb. If she is shipped as a lamb in early gestation, she will command the market price for slaughter lambs. If you wait until after lambing to ship her as a yearling, then she'll get the yearling price, which is ten to twenty dollars less per hundredweight than the lamb price, which means up to twenty-five dollars less per lamb.

Add the feed cost and the loss in market value and you get a thirty- to forty-dollar advantage for pregnancy checking and early shipping of open ewe lambs. Of course, if you want to keep them, that is fine, but be aware of what it costs you. Looking back at the cost of a pregnancy checker of less than three hundred dollars, you can see that as few as eight open ewe lambs can pay for the pregnancy checker, in money saved and earned.

The Lambs

Shepherds who are raising sheep on a once-a-year lambing schedule should be shipping the last lambs from the past season now during the early gestation period. If they are not to market weight yet, they may never get there, or at least they won't make it as lambs and you'll take the losses described in the above paragraphs. As it happens, lightweight, finished lambs are often in short supply in winter, so put your late ones on a lot of free-choice grain, get them finished to grades Choice or Prime (see appendix 2), and ship them. This is no time to fool around feeding lambs on hay. Get them fat and out.

5

LATE GESTATION

Of the approximately twenty-one weeks between breeding and lambing, the last six are the most important. The ewes needed to have a generous food ration immediately after breeding to get the pregnancy off to a good start, but then a maintenance feed level was adequate for almost three months. During that time the fetus or fetuses increased in complexity to the point where most of the final organs were formed by the almost miraculous train of events in which a single cell divides and transforms into the specialized cells of distinct organs and parts of the growing lamb.

During the last six weeks of gestation the already complex fetus grows in size and achieves the final form of a lamb. This rapid growth puts a lot of demands on the ewe. She needs better feed, closer watching, and more detailed attention. A well-orchestrated late gestation will mean a simpler task at lambing, and every bit of effort that the shepherd puts out now will be more than repaid when the ewes start birthing.

NUTRITION

The ewes will require a great deal more food during the last part of gestation. They can no longer be expected to live partly off their own reserves when they are carrying fast-growing lambs. A ewe giving birth to twins or triplets needs protein and other nutrients, not only for the lambs themselves but for the placenta and associated tissues. We have had ewes give birth to triplets with a total weight of forty pounds, not counting the placenta and other tissues, so the demands are very large indeed.

Food Requirements

The total feed given to the ewes must be increased 60 percent or more to satisfy fetal growth. If the early gestation ration has been marginal, an even larger increase may be called for. For an average-sized ewe who has been getting three to four pounds of hay a day it will mean a jump to five to seven pounds of hay, which is getting close to the limit of how much food a ewe can stuff into her rumen.

One feeding method that is used by some sheep raisers is to give the ewes better quality hay during late gestation. The term "better quality" should be examined closely, however. If the hay is the same as that previously fed (except that it is greener and just looks better), then no gain has been made. It may well have a little higher protein and perhaps more carotene to provide vitamin A but still not be nutritious enough. What is needed is a lot higher protein, such as might be provided by a later cutting of alfalfa. If in doubt, have the hay analyzed.

Giving better quality hay may not be sufficient because the ewes still need a 60 percent increase in total digestible nutrients (TDN). However, the rumen can accommodate only so much hay, and while alfalfa cut in the earliest bloom stage may be over 50 percent higher in protein than lower quality hay, it is only about 10 or 15 percent higher in TDN. Thus the ewe might be getting plenty of protein but still be lacking in energy, because she won't be getting enough calories.

The usual solution to this dilemma is to provide a high-energy feed such as grain along with the hay. A ration of two pounds each of corn and hay will amply provide for a ewe's needs and won't fill up her rumen too much. Let me emphasize I am assuming that a lot of the ewe's protein requirement is coming from the roughage, the hay. Corn is only about 10 percent protein. This means that the hay must be about 12 percent protein, at a minimum. I stress this point because I hear of sheep raisers who have tried to carry the ewes through early gestation on straw or corn stalks, then think that they can make up for their past sins by giving the ewes, in addition to the same roughage, a couple of pounds of grain apiece for the last month. The corn or other grain will help with energy needs but will not even come close to providing the protein needed. In such a case, if straw or such feed was the only roughage, a high-protein source such as soybean meal would have to be provided as well as the grain.

If you have been a bit casual about the rations you provide your sheep for the rest of the year, get serious now. Refer to appendix 5 and spend a few minutes calculating whether you are feeding enough of the right things. Incorrect feeding now can ruin a year's planning and work and make for many unhappy hours in the barn with weak lambs, sick ewes, and even abortions caused by malnutrition. If you anticipate a lot of triplets, be sure to calculate the ration accordingly and feed still more because the ewes will need it. A ewe carrying twins doesn't require much more feed than one carrying a single fetus. A ewe carrying triplets has energy needs that are about 15 to 20 percent higher than the ewe with a single or twin fetus.

Mineral needs must be carefully evaluated too. The ewes need calcium and phosphorus and some trace elements in order to do their job of building lambs. Two pounds each of corn and quality alfalfa will provide the minimum needs of calcium and phosphorus. Many producers also give a free-choice mixture of salt and dicalcium phosphate in a half-and-half mixture to ensure that there is adequate calcium and phosphorus. If the calcium content of your hay is low, as is usually the case with grass hays, you would be wise to balance the phosphorus in the grain by adding 2 percent finely ground limestone to the corn or other grain.

Trace elements are usually present in sufficient amounts in ordinary feed, but you can use trace mineral salt in your salt-dical mixture to be safe. Remember that there is a problem of copper poisoning with sheep, so never add trace mineral salt to feed or put molasses in a salt mixture to encourage consumption.

If you have found that selenium is deficient in your feeds, you can give a free-choice sheep mineral mix that contains supplemental selenium. If you prefer to prepare your own mineral mix, there are also selenium premixes available that can be blended with salt and dical to make a satisfactory mineral supplement. Be sure that mixing is thoroughly done, because selenium is highly toxic in excess—so don't just throw in some premix and stir it around with your hand. For the same reason, be sure to check your calculations as to how much premix to use so you don't misplace a decimal point and feed too much or not enough. A salt mixture cannot legally contain more than 90 ppm selenium, which is only .009 percent. If you mix your own, calculate carefully so as not to overdose. One must be sure that sheep have adequate salt available at all times so individuals do not gorge and receive toxic levels of selenium or other trace elements.

Many ewes will need some additional vitamin E at this time as well. Ewes eating hay rather than green pasture will doubtless be short on E. You should supplement either in feed or in a salt/mineral mix at a rate to give the ewes something on the order of 100 IU/day. Ewes generally eat about ¼–½ ounce a day of the salt mixture. Excess E will not be toxic, but not needed either. Do not depend on the E in ADE mixes added to feed. The amount of E in those is trivial, being there primarily as a preservative. The E will not keep until the following year, so plan to use the whole bag and feed accordingly.

Grain Feeders

If you are feeding grain, you have to have some sort of feeder, and some types are better than others. As far as I know, there is no such thing as a perfect feeder. If hay is fed in feeders, perhaps the grain can just be poured over the hay. Mixing of the grain with hay eliminates the problem of a ewe taking too big a mouthful and choking. If feeding grain with hay is not practical for you, then grain feeders of some kind will have to be used.

The first rule of trough design is that the sheep should not be able to stand in it, for the obvious reason of preventing their tracking manure into the feed. This rule eliminates shallow, open troughs. Raising a trough off ground level is some help. To further frustrate the ewe who likes to stand in her feed, you can run a crossbar down the length of the trough. Another practical design is to have narrow slots along the sidewalls of the trough so the sheep have to stick their heads through to eat. Such feeders are generally rather heavy because of all the lumber used and are best for permanent installations, unless they can be moved with a tractor-loader or the like.

In order to reduce the problem of the sheep eating too fast and choking, use a design that forces them to take small mouthfuls. One way to do this is to nail cleats to the bottom of the trough or leave nailheads sticking up an inch or so. One producer told me that he puts small rocks in the trough bottoms to accomplish the same end.

I finally made some feeders that work well, after a number of failed experiments with various types. I obtained some fifty-five-gallon drums made of heavy plastic from a cheese plant. These were cut lengthwise with a saber saw to give four troughs. A two-by-six board was then nailed lengthwise to bridge the space between the two pie-wedge shaped ends, then two-by-four legs were attached to hold the troughs off the ground. I put three of the basic units together to make a three-compartment trough that will feed up to eighteen wooled sheep at a time. These troughs are light and sturdy, and sheep don't stand in them at all. Also, the ewes have to stick their noses under the two-by-six to get a mouthful of grain, and they are inhibited from taking too much at one bite. (The sketch that opens the chapter on flushing shows some of these troughs in use.)

A sheep-raising neighbor got some of the same sort of plastic drums, cut them in half lengthwise, and placed them on the ground with no legs attached. He says the sheep won't stand in them because they roll underfoot. I'd still be a little concerned with the gulping problem, but perhaps a few rocks would take care of that.

Time of Lambing

Late gestation is the period when the shepherd can help determine the time of day that lambs will arrive. The day of lambing is pretty much established by the day of breeding and the ewe's natural gestation period, but the time of day can be influenced. Sheep, like all animals, have natural rhythms for all of their bodily functions. The daily rhythms are termed circadian (from the Latin for "approximately daily"), and they control all daily events in a sheep's life, including the time of day of lambing. Scientists have found that most mammals, including people, have internal clocks of some sort that regulate the circadian rhythms, but they also have found that the internal timing is not exactly on a twenty-four-hour cycle, but may be longer or shorter. However, animals receive clues from external sources during each day that reset the internal clock to synchronize it with a twenty-four-hour day. The so-called jet lag that travelers encounter when they travel long distances east or west in a short time is a consequence of the internal circadian rhythms having to adjust to a sudden change in the time of the external clues, the most important of which is the time of sunrise and sunset.

Unless the shepherd confines the sheep in a light-tight barn, there is no way to alter the effects of the times of sunrise and sunset. What other clues are presented to a sheep each day? I can think of none other than time of feeding, and scientists who are concerned with such things have found that the time of feeding of a hungry animal is indeed a clue that resets internal timing. We have experimented with the time of feeding of grain to our flock and have found that feeding at noon results in about two-thirds of the births occurring between six in the morning and six in the evening, with almost none between ten at night and two in the morning. It is very important not to disturb the sheep at night more than needed. A very low-wattage light, such as a night-light bulb high in a barn, will provide enough light to dark-accustomed eyes so that bright lights do not have to be turned on. Move slowly and quietly and for heavens sake don't have a radio blaring away at night.

Feeding at six in the morning roughly reverses the lambing schedule, with most lambs coming between six in the evening and six in the morning. Thus a shepherd can choose, to some extent at least, when the lambs will arrive, which is quite a convenience.

Some farmers and university researchers have reported that cows fed in the early morning and again at night have mostly daytime births. I have seen opposing reports on lambing of ewes on this schedule. I have never tried this plan. I can report that noon feeding does work, both for me and for a number of shepherds who have tried it and told me about their results.

ENVIRONMENT AND EQUIPMENT

The main change in the environment of the flock is that the expectant mothers should have access to a barn or other shelter for lambing. Shelter may not be a vital thing in some climates insofar as protection from the weather is concerned, but it is important for the convenience of the shepherd. It permits the ewes to be penned up with their lambs for the first two or three days to get acquainted and also keeps them where the shepherd can observe them conveniently.

In cold climates the lambs will profit from some protection from the weather for at least the first couple of days, although once a lamb is dry it can tolerate extremely cold conditions. Shearing of ewes prior to lambing is highly recommended, and the freshly shorn ewes will need housing right after shearing, especially if the weather is cold or wet, until they start to grow a new coat.

Actually, the shelter should be available prior to either shearing or lambing because a ewe that is soaked with rain or covered with snow and ice cannot be shorn. A moist fleece will rot, and drying fleeces off the ewe is not practical. In addition, sheep with wet fleeces carry moisture into the barn where it is not wanted. For these reasons, protection from precipitation is very helpful. An open-sided structure such as a hay shed is appropriate because its function is to keep new precipitation off the ewes while their own body heat and shaking gets rid of any accumulated snow or wetness. Just giving them access to a shelter will not necessarily do the trick, because some of them still stay out in the rain or snow, so they should be confined during precipitation in the days prior to shearing. Even if the ewes are dry, it is worthwhile to pen up a group closely for the night before shearing because the warmth seems to make the sheep easier to shear the next day.

After they are sheared, the ewes will need little encouragement to stay inside in cool or cold weather. This stay-at-home tendency is very helpful at this time because the ewes will lounge in the barn instead of in the yard or pasture and will be much more likely to be inside for lambing—a great convenience for the shepherd-midwife and possibly a lifesaver for the lambs.

A useful plan is to encourage the ewes to stay in the barn or other lambing quarters, but to feed them somewhere else so they still get lots of exercise. It may be just luck, but Teresa and I very rarely experience any of the really bad lamb presentations: the tangled legs of multiple lambs, the head that is back and impossible to straighten, or the other uterine mixups one hears about. We attribute this to the fact that the ewes get plenty of exercise; this keeps their muscles in tone, and the movement helps get the lambs into the right positions in the uterus before lambing begins.

Now is the time to get the barn and equipment in final shape to be ready for lambing. Make sure that all of the panels were repaired to a usable state, replace lightbulbs or wash the cobwebs and flyspots off the ones that still work, and make sure bedding is adequate. Get out your scale and make sure it is in working order to weigh the lambs at birth. The easiest scale is a hanging one. Put the lamb in a big plastic bucket and hang the bucket on the scale hook. Be sure the scale is one that allows you to tare off the weight of the bucket so it reads directly in lamb weight, without mental arithmetic. You can also make a sling out of some canvas and a couple of sticks sewed to each end that holds the lamb under its belly, and can be hung on a scale. Such a sling is also handy for carrying lambs around.

This is also the time to make sure that you have all the equipment and supplies you'll need during lambing, and that you have them where you can get to them when you need them. We kept most of our supplies in a big cabinet on the barn wall that is only five inches deep. With that shallow cabinet, things don't get hidden behind other things, and it doesn't take up a lot of floor space either. Everyone will have a slightly different list of items, but I'll give you a list so you can use it as a starting point for your own compilation.

Most of the items are things you don't need in late gestation, but when the lambs start popping it will be too late to start stocking up. Also, you may well find that you can save a lot of money by ordering many items from mail-order suppliers. In addition, with all due respect to local area merchants, sheep items are often simply not available at local outlets without a long wait for a special order. If you are sufficiently organized to order well ahead of time, you might stir up some interest in sheep products so that next time your local druggist or veterinarian would have them in stock. I'll give a checklist of items and then explain more about them following the listing.

Before Lambing

dexamethasone
coarse twine
examining gloves

mild bar or liquid soap
lubricant
calcium-magnesium solution

At Lambing

iodine solution
towels
navel cord clips
ear syringe
oxygen
Dopram V
blood stopper
twine
hair dryer
heat lamps
lamb nipples and bottles
stomach feeder

oxytocin
colostrum
lamb milk replacer
5 percent dextrose solution
selenium injectable
penicillin
Cl. perfringens type C-D antitoxin
PI-3 nasal vaccine
injectable iron-dextran solution
electrolyte mix
uterine boluses

After Lambing

docker	Terramycin
Elastrator tool and bands	soluble sore mouth vaccine
tetanus antitoxin	*Cl. perfringens* type D antitoxin
Quartermaster teat syringe	kaopectate

Dexamethasone (Dexasone, Azium) is one of a class of compounds called corticosteroids, all of which have potent physiological effects. It can be used to induce lambing if that is needed. It and similar compounds (such as flumethasone) are available only from your veterinarian and should be used under a vet's guidance. I'll discuss its use under ketosis in this chapter.

Coarse twine is useful to help pull out lambs in some difficult births. A woven tape called umbilical tape that you can get from your vet is equally good or better, or a gauze bandage may be used. Smooth cord such as that made from cotton or nylon is too slippery to be useful. Cut some of the twine, gauze, or tape into three-foot lengths and store it in a jar in some rubbing alcohol diluted one to one with water.

Examining gloves are thin rubber gloves that can be slipped on before entering the ewe with your hand. They come only up to your wrist, but they cover dirty fingernails. You can buy these from a vet or physician. Discard them after one use. They also afford protection to you against infections from an infected uterus.

Soap is available as mild bar soap without perfume or antibiotics or as liquid green soap at your drugstore. This is used to wash up before entering a ewe and to clean up afterwards. It can be used as a lubricant as well.

A lubricant is helpful to make the lamb slip out more easily or to let your hand and arm slip in more easily. Some people just use soap, which is fine provided it is not an irritating sort. My favorite lubricant is one sold by the McGrath Company (P.O. Box 148, McCook, Nebraska 69001) called lubricant gel. It is a powder that is mixed with water to produce a nonirritating, thick liquid that is unbelievably slippery. Liquid soaps are positively raspy by comparison. McGrath sells mostly to vets, but will sell direct to you if you don't have a local source.

Solutions containing calcium, magnesium, phosphorus, and other materials are used for the treatment of milk fever, a condition that can affect ewes before or after lambing. These solutions are sold under brand names like Caldex, Cal-Phos, and Cal-Mag. I'll discuss their use under Milk Fever in this chapter.

Iodine solution is used to apply to the navels of newborn lambs. Don't waste money on spray containers of iodine. You can buy it in gallon jugs much more cheaply. You can use either the so-called gentle iodine or the 7 percent tincture. The 7 percent type is more irritating, but it also has a stronger drying action on the umbilical cord.

Towels are a requirement at lambing as far as I am concerned. In cold weather you'll want to dry the lambs and your hands. We have a collection of old terry towels that we have scrounged through the years. Try friends, relatives, and garage sales.

Navel cord clips are little plastic gadgets that are clipped onto the umbilical cord of the lamb at birth. They are supposed to prevent disease-causing organisms from entering the lamb's body via the broken cord. We used them some years and not others with no apparent difference in lamb health. Now and then a lamb will bleed profusely from the broken cord, in which case a clip can literally be a lifesaver, so having a few in reserve is a good idea. We keep some in a jar of rubbing alcohol to be ready for use if needed. You will find that ewes chew them off and they get lost in bedding, so have extras.

An ear syringe such as those for human babies is used by some to clear the lamb's nostrils. A towel and finger do just as good a job in most instances. An ear syringe can also be used to administer a soapy water enema.

A small tank of breathing oxygen is a handy luxury in the lambing barn. We use a portable tank that is intended for use by a light-plane pilot. The face mask that was intended for a human fits a lamb just right if it is held upside down over the lamb's nose and mouth. A short (15- to 20-second) shot of oxygen will do wonders for a lamb that has had a prolonged and difficult birth. Oxygen is not essential but can be very useful. If you cannot find a used bargain at a local airport, you can rent small tanks from medical rental places.

Dopram V (Doxapram Hydrochloride) is a potent drug that stimulates breathing. It can be a lifesaver for slow, weak lambs. You'll need to get it from a vet. I'll discuss in the next chapter.

Blood stopper is a powder that is applied to bleeding wounds to staunch the flow of blood. If your barn is sufficiently rustic, you can grab a handful of spiderwebs and apply them to the bleeding to help coagulation.

A few lengths of twine are helpful for tying a lamb's legs together to help in grafting him onto a foster mother as I'll describe later in the chapter on lambing. Some producers even tie twins together by one leg so they don't get separated in the crowd with one of them becoming abandoned. Twine is also indispensable in treating a prolapsed vagina, as discussed later in this chapter. Twines saved from hay bales are fine.

A hand-held hair dryer is a wonderful tool for lambing and can be used to dry hair during other parts of the year. There is nothing better for drying off a lamb quickly, which is very important in cold weather. The dryer is great for keeping a lamb's ears from freezing off. Dry ears are quite safe from freezing, but wet ones freeze right away. One caution: Some hair dryers really put out the heat, so be careful that you don't burn the lamb.

Heat lamps are useful to help slow lambs get going by getting them warmed up enough to be interested in nursing. Heat lamps should be used in moderation, and some shepherds think they should not be used at all. They certainly should be used only enough to get a lamb started, and that's all. If heat lamps are left on for long periods they seem to predispose the lamb to pneumonia, and there is no doubt that they are a fire hazard. Our fire inspector insisted they be hung from chains rather than rope or twine.

Some Pritchard lamb nipples and a few screw-top pop bottles are necessary bits of equipment. The holes in the nipples can be enlarged with a hot needle to increase the flow of milk.

A gadget of some sort for stomach feeding of lambs is a must if you want to save most or all of your lambs. I'll discuss the various alternatives in Lambing.

Oxytocin is a hormone that is present in any lactating animal. It causes the so-called letdown of milk in the udder. Natural production of oxytocin is stimulated by the hitting and sucking that nursing lambs do; it can also be stimulated by massaging of the udder. Some ewes do not let their milk down at first, presumably because some hormonal message is late. In such cases the shepherd can effect letdown by giving an injection of oxytocin. This is an item to get from your veterinarian.

Colostrum is the first fluid that comes from a ewe's udder when she freshens at lambing. It is not just rich milk; it also contains special proteins called antibodies that give the lamb temporary protection from disease-causing organisms. It is important that every lamb get some colostrum as its first meal, and you should have some on hand in your freezer for those lambs that get none from their mother for some reason. Ideally, you should have some sheep colostrum, but goat or cow colostrum will do. Steal some from each ewe and husband it for future use. The frozen stuff can be kept for years in airtight containers. Let it thaw slowly, without heating, because temperatures above 125° F will destroy some of the antibody protein.

You should have some lamb milk replacer on hand for orphan lambs. This is not a substitute for colostrum, and the lamb should get some colostrum before getting any other food.

A bottle of 5 percent dextrose (or glucose) solution is useful to provide a quick shot of energy to weak lambs. Some producers prefer to use solutions that also contain amino acids and electrolytes. Avoid the 50 percent solution since it has a dehydrating action. However, the 50 percent can be mixed with an equal amount of boiling water to make a warm 25 percent solution for injection.

There are injectable selenium and vitamin E mixtures that are used for prevention of deficiencies of these nutrients. If you have experienced white-muscle disease with lambs, you may want to inject newborns with this preparation. This is available from your veterinarian. Use only the product intended for lambs, not the stronger one for adult sheep. If you provide sufficient selenium in salt, this should not be necessary.

A bottle of penicillin G (the cheapest form) is handy at lambing. Some veterinarians recommend an injection of penicillin to all newborn lambs as a precaution against infection. One can give a penicillin injection to each lamb that was born unattended and hence did not get the navel cord dipped in iodine solution at birth.

If a lamb did not receive colostrum, it should get an injection of antitoxin against *Cl. perfringens* types C and D. The type C toxin is a common cause of lamb scours.

PI-3 nasal spray vaccine can be given to all lambs at birth as a precautionary measure against pneumonia. The use of this vaccine in sheep and lambs is not approved by the FDA.

It is commonly said that lambs do not suffer from iron deficiency. As I will mention in the next chapter, that may not be so. Injectable iron dextran such as is used with baby pigs can be given to lambs if you wish.

Electrolytes are soluble materials such as ions of sodium, magnesium, potassium, bicarbonate, and other chemicals. A lamb with scours loses fluids and electrolytes. Electrolyte solutions given orally (by a bottle if the lamb will suck or by stomach feeding if not) help to replace both lost water and electrolytes. Many commercial mixtures are available. A home-brewed electrolyte and amino acid mixture recommended by Colorado State University is as follows: 1 package fruit pectin, 1 teaspoon Morton Lite Salt, 2 teaspoons baking soda, 1 10½-ounce can beef consommé, plus warm water to make two quarts of solution. Make up a fresh batch as needed. Do not use a substitute for the Morton Lite Salt, as it is used to provide potassium in the correct amount. Gatorade can be used, but it is inferior to the above mix.

Uterine boluses are placed in the uterus of the ewe after the shepherd's hand has entered, since it may carry with it pathogenic bacteria or other microorganisms. Our veterinarian thinks they are completely useless. Decide for yourself or ask your vet for a second opinion.

Ear tags come in a variety of sizes and shapes. In order to keep meaningful records, you must have some way of identifying each lamb. I discuss the various types of tags in chapter 6. Be sure to get an applicator for whatever type you decide to use.

Many sheep raisers paint-brand ewes and lambs in order to be able to spot quickly who belongs with whom. Paint brands are no substitute for ear tags because many brands become unreadable soon after application. Paint brands are handy, but you'll have to decide whether you think the extra work is justified.

You will need some way to dock (remove) tails. There is an electrically heated device that is highly satisfactory, and there are propane-heated ones also. Other gadgets that crush the tail to reduce bleeding are used with success too. Some shepherds use Elastrator bands.

Castration can be done with some docking tools or with various special tools. More information about both docking and castration is given in the next chapter.

To treat inturned eyelids (entropion) in lambs, have a syringe of Quartermaster mastitis treatment on hand to use, as described in chapter 6.

A dose of soluble Terramycin stops scours in some lambs. The problem with scours is that different flocks have different resident microorganisms, so what works for one may be useless for another. Try various antibiotics until you find one that works. Asking the person from whom you bought the sheep might be helpful. Mechanical aids such as Kaopectate are used by some.

If you are going to vaccinate for sore mouth (ovine ecthyma), you'll have to have the vaccine in stock. You will also be vaccinating all lambs against *Cl. perfringens* types C and D at some time, so order enough of that toxoid or bacterin to have on hand at weaning.

One final preparation that you may want to consider is to arrange for hired help to assist at lambing. This is especially important if one or more of the family holds an outside job and cannot be in the barn for part of the time. The way to save lambs is to have someone present for every birth. Nobody can be there all of the time, so get some help if it is needed. Lambing commonly occurs during a slack period on farms, and you may have neighbors who would welcome a part-time job at that time.

HANDLING: SHEARING

Anyone who has lambed both shorn and unshorn ewes will tell you that you should shear before lambing, at all costs. There are a host of good reasons to shear before lambing. The ewes take up less space in barns and pens when shorn. They also will carry less moisture in their wool, and the wool will not get contaminated with bedding in the barn during and after lambing. A shorn ewe will stay in the barn at lambing (a convenience for the shepherd) and will also tend to stay in the barn in cold and wet weather after lambing so the new lambs are less likely to be exposed to weather that they can't tolerate. In addition, a shorn ewe is much easier for the shepherd to evaluate prior to lambing. Her vulva and bag are easily visible, as are her flanks, so the shepherd can see changes in her body conformation that precede lambing. Furthermore, a cleanly shorn underside makes it easier for the lamb to find the teats. On an unshorn ewe, some lambs will suck on bits of wool or dung locks instead of teats, and there is very little nutrition in a dung lock except if you put it in your garden.

From the point of view of a shearer, a pregnant ewe is a delight to shear. The swollen abdomen stretches out all of the wrinkles and makes the job easier, not to mention the fact that a pregnant ewe is a lot less feisty than a slender, open one. The quality of the wool may be better preserved by timely shearing before lambing. Some ewes will develop a so-called break in the wool at lambing. A break is a place where the fiber diameter of the wool is greatly reduced, so that it is weak at that point and will break when stretched. If the wool is shorn right at the break, there is no adverse effect because the break is at the end of the fiber rather than somewhere in the middle. Breaks in the wool can be caused by sickness, but the break at lambing is most likely caused by diversion of protein from wool growing to the growth of the lamb as well as to the production of antibodies for the colostrum.

If for some reason it is impossible to shear the ewes prior to lambing, they should at least be crutched to expose the vulva and trimmed around the bag area. Do try to get them totally sheared, though, if you possibly can.

In some areas shearers are hard to find, especially for the small flock. If you have difficulty finding a shearer, or are unhappy with the job done, then by all means learn to do the shearing yourself. Shearing isn't difficult; it just takes practice. In order to learn, you can work with a shearer or try to attend one of the shearing schools that are often sponsored by local and state sheep associations at members' farms and at universities with agriculture programs. Check with your county extension agent to locate the nearest school, or ask on an Internet sheep list. There is a widely praised shearing video entitled *Shearing Techniques with Fiona Nettleton* available from Rural Route Videos, P.O. Box 359, Austin, Manitoba, Canada R0H 0C0, 800-823-7703.

Don't be nervous and shy about learning to shear. Everyone at a shearing school was a beginner once, and they will go out of their way to give you pointers and to help you. If you have only a few sheep to practice on at home, try to locate some others to practice on, because only by shearing a lot of sheep can you become better. After you have shorn a hundred sheep or so you should be able to do one in less than ten minutes, maybe a lot less. If you are going to try to shear a lot of sheep at one time, be sure to have a helper or two to catch the sheep and set them up for you, and to take care of the shorn fleece. If you will catch for a good shearer you can see how a professional goes about it. He'll probably let you do one yourself now and then and will give you some valuable tips. Hardy's Gabriel Oak in *Far from the Madding Crowd* sheared a sheep in record time with hand shears.

"Well done, and done quickly!" said Bathsheba, looking at her watch as the last snip resounded.

"How long, miss?" said Gabriel, wiping his brow.

"Three-and-twenty minutes and a half since you took the first lock from its forehead. It is the first time that I have ever seen one done in less than half an hour."

The clean, sleek creature arose from its fleece—how perfectly like Aphrodite rising from the foam should have been seen to be realized—looking startled and shy at the loss of its garment, which lay on the floor in one soft cloud, united throughout, the portion visible being the inner surface only, which, never before exposed, was white as snow, and without flaw or blemish of the minutest kind.

—THOMAS HARDY, *Far from the Madding Crowd*

Shearing equipment is of three types. The traditional hand shears are rarely used today to shear a whole sheep, although there are purists who still depend on their own hand power. Every sheep raiser should have a pair of hand shears for trimming and for fitting sheep for show and other light-duty jobs, but serious shearers should have a motor-driven shearing machine.

The least expensive power-driven shearing equipment is a hand-held electric motor with a sheep shearing head attached. For the small flock or for the beginner, that type is a useful tool and is very handy to have around even if more elaborate and powerful equipment is used for big jobs. The shearing head holds the cutting parts, which are called the comb and the cutter. The comb is a flat piece of metal with long teeth that penetrate the wool readily. Another metal piece, the cutter, oscillates back and forth over the comb to give the cutting action. Both comb and cutter have to be sharpened on a special sharpening plate or can be sent off for custom sharpening if the small producer doesn't want to invest in a sharpening rig. Lister, Oster, and Australian Sunbeam are common brands.

The self-contained type will do the job, but it is less powerful than larger machines. At a shearing school you will be taught how to adjust cutter tension properly and how to change combs and cutters and align them. Even if you are shearing only a few sheep, you should have three or four sets of combs and cutters.

For speedier shearing with large flocks you will want to invest in a large shearing apparatus. The big machines use a remote source of power, usually an electric motor that is hung on the wall or on a post. Power is transmitted through a jointed or flexible shaft to a handpiece that holds the comb and cutter. Because there is no motor in the handpiece, it is smaller and easier to use and doesn't get overheated, as the small motors can in heavy use. The extra power and wider cut possible with a remote-powered handpiece makes it the choice of the pros. The only difficulty for a beginner with a remote-powered unit is that the sheep has to be brought to the shearing machine rather than the other way around. Once a shearer gets enough experience, the sheep never has to be moved from an area about two feet square, but a newcomer may use an area the size of a small room.

Perhaps the best advice for a beginner is to learn with a self-powered device and then graduate to a bigger machine when and if skills and sheep numbers make it appropriate. Many shepherds will never need a bigger machine than the small ones. Incidentally, the small machines can be equipped with a clipping head that is useful for clipping cows, goats, dogs, or other animals; it also does fine trimming on a sheep when fitting for showing.

Many shearers use a general-purpose nine-tooth comb, but there are variations for special purposes. Many combs flare outward to gather more wool in a single pass. A nine-tooth thin comb is a good compromise, giving a less tidy appearance than a thirteen-tooth one but being faster and easier to use. If the shorn sheep will not have access to shelter, or in extremely cold weather, one can use nine-tooth protector combs that have runners under some of the teeth to cut a bit farther from the skin and leave a stubble of protection for the ewe. Some shearers don't like these because they are slower, but insist on them if you have special needs—recognizing that you may have to pay a little more.

A self-powered machine with one comb and cutter set will cost up to $200. The handpiece alone of a larger setup can cost twice that. You can expect to pay another $300 to $500 for the rest of the machine. A sharpening rig will cost about $150 plus the cost of a motor. Combs and cutters cost $10 to $30 and up depending on the type.

Whether you or someone else does the shearing, the barn or shed should be prepared ahead of time. Do not bed the area down with straw just to make it look nice, because the straw will get into the wool and reduce its value. If the area needs bedding to cover mud and droppings, do the bedding a week or so ahead of time to give the sheep a chance to stomp it down a bit. You will need a place to accumulate the wool as it is sheared. If you sell individual fleeces to handspinners, you can roll up each one and put it in a plastic garbage or leaf bag. Do not close the plastic bag, because moisture has to be able to escape. If you sell your wool to the commercial market, you should pack it, untied, in a standard wool bag. Wool bags are giant burlap sacks that are hung from a sturdy stand so that the bottom of the bag does not touch the floor. They hold about two hundred pounds and up and are filled by a person standing in the bag using his feet to stomp and pack the wool tightly. The current trend is to bale wool and not use bags that contaminate the wool with vegetable or plastic fibers. The small flock owner doesn't produce enough wool to make bales, so check with local buyers to see what they require.

You will also need a shearing area that has some sort of smooth floor that can be swept clean between each shearing. A wooden floor is excellent, or you can use a large piece of heavy carpet turned upside down so that you are shearing on the woven back. If you live near a paper mill, try to get a piece of the very wide belting they use when belts are replaced. The belts are ten to twenty feet wide and made of very heavy material. Don't try to shear on a tarp or blanket because these bunch up and are more of a nuisance than a help.

As you finish a sheep, the helper should tie the fleece with paper twine made especially for that purpose and place it in a bag or in a holding area. Never use any other kind of twine to tie a fleece, because fibers from the twine can contaminate the fleece, and a wool buyer will dock you or refuse to buy the fleeces. Paper twine is used because it softens in water and disintegrates so that it can be washed away with the dirt from the fleece. The modern trend is not to tie the fleeces. Again, check locally or ask your shearer.

If you use plastic twine, your wool will almost certainly be rejected by a buyer, because there is no practical way to remove plastic fibers from wool. For this reason, never use hay that is baled with plastic twine. The short plastic fibers left when the baler cuts the knot will get into the wool and totally ruin it.

Separate your wool into different types as you pack it. Lamb's wool should be separated from adult wool, and the adult wool should be separated according to type if you have enough to do so. You will get a better price for wool that is grouped.

Shorn sheep can be dusted with an insecticide powder to kill keds and lice, and hooves can be trimmed if needed. Be alert for bloat, because the excitement of shearing sometimes brings it on. Shearing also is stressful because usually strangers are around.

After shearing, the wool should be properly stored until it is sold, which usually means a place where it is kept dry. Wet wool will rot and be worth nothing. Wool bags stored on a damp floor invite trouble because the bag itself may rot and rip when you try to move it. Too much of a good thing is never right either, and storage in a very dry place will result in a big loss of moisture from the wool, which in turn means a loss of weight. Since wool is sold by the pound, that lost water is money out of your pocket. Wool that is stored on the ground floor of an unheated building will retain a suitable amount of natural moisture, especially if the pile of wool bags is covered with a tarp to keep dust off. Never store wool in a heated building in winter unless you are willing to accept weight losses of 20 percent and up. Also, keep the larceny in your soul under control and don't be tempted to turn a hose on the bags or leave them in the rain on the day before selling. I know a fellow who does that sort of thing, and he has the respect of nobody.

If stored properly, wool can be kept for a long time, and many people stash it away to wait for the right price. If you do this, remember that wool in storage gathers no interest, so don't let your stubbornness cheat you out of the best total price for your wool. If you are selling wool to handspinners, never keep it for longer than six to eight months. The so-called yolk in the wool (a combination of lanolin, sweat, and dirt) will gradually harden with time until the wool is very difficult to spin without first washing it. Many spinners prefer to spin unwashed wool—this is called spinning in the grease—and your wool will not be satisfactory if it is old and dried out. Sell old fleeces along with your commercial wool, and don't ruin your reputation by trying to palm them off on a handspinner. The current trend among handspinners is to wash wool before spinning, or even to send it to a professional for processing. You may want to investigate having your wool processed before you sell it.

During shearing, be sure to handle the ewes gently and avoid crowding them. We have sheared ewes on the day before they lambed with no problems, but be easy with them. If you use chutes, be sure they are just wide enough for one ewe so that two don't try to crowd side by side. Avoid moving the ewes through narrow doorways or gates, and avoid having the flock make a turn just after they go through a gate or a doorway. The flock will always crowd toward the inside of the turn: pity the poor ewe who is caught against a fence post or door jamb. If a turn is unavoidable, post a person or a dog at the inside of the turn so the flock will shy away a bit. Avoid using a dog to herd ewes in late gestation unless the dog is trained to hang well back and not crowd the flock together. Be sure there is plenty of space at feeders. A feeder that gave plenty of room during early gestation will be mighty crowded when all of the girls are full of twins and triplets, although shearing does help the space problem considerably.

Cautions: If you have a flock of mixed white and colored sheep, be sure to shear the white ones first, so you don't contaminate the white fleeces with black fibers. For sanitary reasons, be sure to try to avoid shearing any sheep with abscesses until last. If an abscess is cut open, do not use the cob and cutter again until it is washed with lots of soap and water. Otherwise you could spread CL through cuts from shearing. See chapter 6 for some information about CL.

Chemical Shearing

There is a product used in Australia called Bioclip that is a synthetic version of the natural chemical that causes sheep to shed their wool. This shearing without shearing sounds appealing, but it has its complications, not the least of which is that the chemical is not approved for use in the United States. To use Bioclip the sheep must first have all dirty and stained parts removed by conventional shearing, then it is enclosed in a big "sweater" to keep the fleece together, then the chemical is given, and soon after the wool starts to come loose from the sheep. The sweater and wool cannot be removed immediately, because the sheep is totally without wool cover, and would suffer exposure, sunburn, insect attack and the like. This "easy" way is not as easy as it sounds. Not only that, it is suitable only for very fine-woolled sheep.

MEDICAL

Late gestation is the time to give the ewes a booster vaccination against *Clostridium perfringens* types C and D. This will be the second vaccination for replacement ewes and the annual booster for the older residents of the flock. The booster increases the number of antibodies in the ewe's blood that are specific against *Cl. perfringens* and the toxins it produces. A veterinarian would say that the titer of the antibody was increased. The reason that the titer should be increased now is not for the ewe but for the lamb-to-be. Antibodies are large protein molecules that are able to attach themselves to bacteria and toxins produced by bacteria and by so doing render the toxins harmless and the bacteria easier for defensive cells to destroy. The vaccination causes the ewe to produce an antibody that is tailor-made to defend against *Cl. perfringens*. This antibody is then passed to the lamb through the colostrum and gives the lamb temporary immunity.

For exactly the same reason, to increase antibody titers, ewes can be given booster vaccinations against other pathogenic organisms that might adversely affect lambs. Examples are vaccines against tetanus, sore mouth, or parainfluenza-3 (PI-3) virus. This is not to say that every shepherd should vaccinate against these diseases.

Tetanus is a rare problem in some areas and with certain management practices. Sore-mouth virus is not present in some flocks and should not be introduced. If you do vaccinate against sore mouth at this time, do it only if all ewes are immune, either from previous vaccination or from having had the disease. The reason for this is that an unprotected ewe will get the disease, and the virus will be spread around the area from the scab formed at the vaccination site, so there will likely be lots of infective material around to transmit the disease to the lambs later. On the other hand, you would like to raise the titer of sore-mouth-specific antibody in the ewe's colostrum to protect the lambs. As you can see, you are caught between the proverbial rock and a hard place. There is no simple answer to this dilemma: you might want to talk to your vet or toss a coin, or both. The goal is not to vaccinate once all sheep are immune, then lambs are protected by colostrum.

There is no treatment that will cure sore mouth. However, on the basis of a comment by animal nutritionist Dr. William MacDonald of California, there have been a number of informal trials of using injectable vitamin B12 applied topically to the scabs (lesions), including one done by a veterinarian with several hundreds of lambs in Mexico. The treatment is very effective. It does not shorten the course of the disease, but the scabs and discomfort are greatly reduced, and the possibility of secondary infection is cut drastically.

If you have experienced pneumonia in lambs during previous lambings, vaccination against the PI-3 virus is probably in order. Vaccination of previously unvaccinated ewes with the injectable vaccine is not appropriate because it causes a general infection or viremia and can cause abortion. The nasal vaccine, especially with previously vaccinated ewes, seems to pose little or no danger. The PI-3 vaccine is not of proven value with sheep, but clinical tests by practicing veterinarians suggest that it is of some value.

However, don't get vaccination-happy and spend a lot of time and money trying to immunize your flock against every known disease. Vaccines are not without dangerous—even lethal—side effects under some conditions. Discuss your whole vaccination and preventive health program with your veterinarian to devise a policy that best fits your flock.

If the ewes were not wormed at mid-gestation, they should be wormed at this time. The previously dormant worms become very active just before and after lambing and will undergo a rapid population expansion if not treated; this will result in poor ewe health and transmission of large numbers of worms to the lambs as well. According to Rupert Herd of Ohio State University, worming is most effective if done in the two weeks just prior to lambing, although it is almost as effective when done immediately after the birth of the lambs, especially if the ewe is penned away from a pasture for twenty-four hours or more.

FLOCK OBSERVATIONS

Late gestation is the time when flock observations are very important. During some other parts of the ewe's cycle you may be able to get away without watching the flock frequently, but not now.

Abortions

Look carefully for ewes who are ready to abort or who have done so. Send fetuses and tissue to a lab for examination, and cross your fingers. If a number of abortions occur, you should react with some sort of treatment. If vibriosis is a possibility you can feed 250 mg of Aureomycin crumbles per head per day in the feed when an outbreak starts. Some causes of abortions such as enzootic abortion or listeriosis are untreatable at present. Be sure the flock is adequately fed, because malnutrition can cause abortions too.

Prolapses

As the ewes become more and more filled by the enlarging lamb, placenta, and other tissues and fluids, there is less and less space for everything. As a result parts of the vagina or rectum or both may push out (prolapse); this is sometimes called eversion of the vagina or rectum. This condition is serious because the exposed tissues can become dirty, frozen, cut, sunburned, and infected. Continued exposure can result in death of the lambs and ewe. The only solution is to try to get the exposed tissues back in place and hold them there. If the prolapse only appears when the ewe sits or lies in certain positions, but goes back in when she stands, treatment may be put off, but watch her carefully.

If a big, red, grapefruit-sized mass appears and stays out, something must be done. There are little plastic gadgets on the market called ewe-bearing retainers that are inserted into the vagina after the tissue is replaced. The retainer is is positioned on the cervix and held in place by tying it to the wool or by stitching it to the skin of a shorn ewe. Some people claim to have good luck with these or with homemade versions, but I think that they are a waste of time for all but the least serious cases of vaginal prolapse. The reason they don't work is that the ewe strains and pushes and flips out the vagina again, plastic gadget and all. For similar reasons, stitching the vulva closed to keep everything in is not very satisfactory.

The method we like is one suggested by Joseph Rook, extension veterinarian at Michigan State University. It involves making a harness for the ewe. You'll need four or five pieces of baler twine tied together into one long continuous piece (see, there *is* a use for old baler twine!). Place the middle of the twine over the top of the ewe's neck and let the two halves hang down on each side. Cross the two ends over the brisket. Then bring them up under the two "armpits" and up the flanks. Cross them again over the spine just rearward of the last rib and then run them under the rear "legpits," between the thigh and the udder. Then bring them up and over the rear end, one twine on each side of the vulva and tail, and then forward to the original twine over the back of the neck, looping the two free ends under the original piece and tying them temporarily. Clean the prolapsed tissue with warm water and mild soap and push it back in place. It helps to elevate the ewe's hind end when doing this. Have a helper hold her hind legs up on a bale of straw or face her downhill. If the ewe pushes the tissue back out almost as soon as you replace it, have a helper hold onto her tongue using a towel or a cotton glove to get a good grip, and pull on it. When her tongue is pulled, she can't strain, at least not as much.

If you are alone, this procedure can still be followed. Loop a double length of twine over your neck and tie the ends to the ewe's two rear hooves. Now stand up. This lifts the ewe's hind feet off the ground while she carries most of her weight on her front legs. (If she rolls over on her spine, then you have to carry her weight on your neck, which isn't nearly as easy.) With her rump in the air, the prolapsed mass will generally pop back in of its own accord.

If all else fails, pour granulated sugar over the moist mass; this will help to shrink it down to the point where it may fit in more easily.

Assuming that you have gotten the vagina back in by some method, now back to the harness. Tie some short pieces of twine between the two strands that pass either side of the vulva, one at the top of the vulva and one at the bottom. You also can tie a piece across the top of the tail. Then the whole harness is tightened by taking up slack where the two free ends were temporarily tied to the piece across the neck. The harness should be so tight that the ewe can barely walk. She may even fall over sideways at first, but set her up on her feet, and she'll learn to walk with it on after a few minutes. The idea of this harness is that the short twines across the vulva keep the tissues in place, and the tight harness prevents the ewe from humping her back and straining. Putting on the harness is much more complicated to read about than it is to do, and you will find that you can do it quickly, even alone, after a couple of tries. The harness really works, and most other methods don't.

Small lambs can be delivered right through the harness. For larger ones the cross-twines can be cut or moved and the harness loosened somewhat. Needless to say, the shepherd should make every effort to be present at lambing of a trussed-up ewe. Some ewes will have vaginal prolapses as lambs or yearlings, then never have one again as they grow bigger and have more room inside. An older ewe that prolapses or one that does it twice should be culled. The tendency to prolapse seems to be heritable so don't keep replacements out of ewes who have prolapsed.

Some ewes bred as lambs may be so small that they cannot make room for twins or triplets and may prolapse both the vagina and the rectum. Harnessing is helpful, but the best solution is to get the lambs out of the ewe as soon as possible. If the ewe is valuable, or if her lambs are expected to be of greater than average value, it may be economically reasonable to take the lambs by Cesarean (C) section. Check with your veterinarian as to cost because the charge for a C-section seems to range widely from vet to vet.

If the ewe is due to lamb within less than four days, lambing can be induced by use of appropriate chemicals. We have used dexamethasone for inducing labor with good results. An injection of 4 cc is given, and the ewe should lamb within thirty-six hours. If she doesn't lamb, repeat doses can be given every twenty-four hours until she does lamb. Other chemicals are also used, such as flumethasone at the rate of 2 mg per injection, according to Dr. A. L. Slyter of South Dakota State University. Be sure to consult with your veterinarian before using these biologically potent compounds, because you might create more problems than you solve. One important consideration is that both of these compounds suppress the ewe's immune system, so she is less competent to fight off infections.

Acidosis

Various digestive and metabolic disorders can crop up during late gestation. Some ewes, especially older ones, may have trouble adjusting to the increased amount of grain in their diets and may respond by getting acidosis. Treat them with baking soda and diet restrictions, as described in the chapter on flushing. I don't mean to pass over acidosis lightly, as it can be fatal. Watch any ewe who is off her feed after having eaten heavily the day before and any ewe who got into some extra feed somehow.

Ketosis

Another disease that can affect ewes during late gestation is ketosis, sometimes called pregnancy toxemia or twin lamb disease, the latter name coming from the fact that ewes carrying more than one lamb are most often afflicted. The usual first symptom is going off feed, as with acidosis, but the similarity ends there. As the disease progresses, the ewe will stagger, look glassy eyed, may circle, lean her head against objects like fence posts, and may elevate her head. Then she becomes weak, lies down on her chest, becomes comatose, and dies. Thin or fat ewes are the ones most usually affected. The breath smells of ketones, like nail polish remover. The presence of ketones in the urine is also a symptom and can be detected using test papers or powders that your veterinarian can supply to you. Collecting a urine sample is easy. Have someone hold the ewe's nostrils closed to prevent breathing. Her reaction usually is to urinate, so be ready to collect a sample.

Ketosis will rarely affect a ewe that is being fed adequately during late pregnancy because the disease results from mobilization of the ewe's own body fat to provide energy for the growth of the fetus. The ketones that poison the ewe are a harmful by-product of the fat mobilization. Some fat ewes will develop ketosis even with adequate feed so it is very important that ewes enter late gestation in good condition without being obese. This is one of the reasons to let the ewes lose some weight during early gestation so that they can be on a weight-gaining feed level when rapid fetal growth demands its share of nutrition.

Even if the shepherd provides what should be enough nutrition, some ewes likely will come down with ketosis. They will usually be those carrying triplets or who don't compete well at the feed trough. Watch for early signs and isolate any ewes who are acting distracted or staggering a bit. The urine test is not completely foolproof, but it can catch some early cases. Treatment of ketosis is usually not very effective. There are gallon bottles of so-called ketosis cures that consist of propylene glycol to be administered as a drench. The energy provided by this material will relieve symptoms for a while, but the symptoms recur as soon as the propylene glycol is metabolized because the unborn lambs are still demanding nutrition. Glucose can be administered as a solution injected intravenously or subcutaneously along with the first drench of the propylene glycol. Be sure not to use ethylene glycol, which is automotive antifreeze and is very toxic.

The only cure once the disease has progressed very far is removal of the lambs. In the early stages, and if the ewe is near full term, the lambing can be chemically induced. If the ewe has already gone down, there is not time to wait for the inducing to work and only a Cesarean-section removal of the lambs will save the ewe. With luck, ewe and lambs can be saved, though of course all can be lost too. You'll have to judge whether the ewe and lambs are worth the cost of the vet's doing the job. Unless the ewe is too far gone, she will recover promptly when the lambs are removed. Consider that the lambs may well be orphans even if the ewe recovers, because the ewe may have no milk or may reject the lambs. In any case, be prepared to give the lambs care, because the veterinarian will be busy with the surgery. Have some colostrum ready to give the newborns for their first meal.

If the ewe is in terminal stages, you may also choose to take the lambs yourself. Make an incision down the midline, locate the uterus, and get the lambs out any way you can. Some people say they shoot the ewe before making the incision. This gives the lambs less of a chance, in my opinion, and I'd opt to get the lambs out while the ewe is still alive. If you are sacrificing the ewe, she is presumably comatose and won't feel a thing anyway. We have attempted this emergency operation only once, and the ewe died before we could get a scalpel so we lost the lambs. It is a grim business, especially with a favorite, but after all you *are* trying to save her lambs.

Milk Fever

Another metabolic disease that can strike around lambing time is milk fever or hypocalcemia, a familiar and worrisome condition to dairy people. The name "milk fever" is a misnomer because fever is not a symptom. Rather, the ewe becomes stiff and uncoordinated, sometimes with a spraddled stance. Tremors, weakness, wide "worried" eyes, and rapid breathing follow. Then the ewe may go down with head forward and legs back. Coma, paralysis, and death soon follow.

The cause of milk fever is a low calcium level in the blood. This does not mean that the diet has insufficient calcium. In fact, high calcium in the diet can actually lower blood calcium levels. Treatment consists of intravenous injection of calcium in the form of calcium gluconate, usually with some magnesium and glucose and other things added. This IV calcium should be administered by a veterinarian until you learn the technique because a too fast administration can cause a heart attack and kill the patient. If milk fever is a common problem, ask your veterinarian to teach you how to monitor the heartbeat and administer the IV solution correctly. The calcium solution can be given subcute in a number of sites, about 10 cc per site. Alternately, the whole dose, usually a rather large amount, can be given intraperitoneally. A needle is inserted into the center of the triangular area bounded by the spine, the last rib, and the hipbone on the right side of the animal. The needle is inserted slowly so that it punctures the skin and enters the peritoneum but pushes the intestine out of the way. Get your vet to show you this triangular hollow so you really will know what you are doing. Either the subcute or the intraperitoneal (IP) injection is much less desirable than the IV because the solution is very irritating to the tissues. The causes of milk fever are not really well understood. Happily, it is not at all common in sheep compared to dairy cows. Keeping the sheep supplied with adequate salt seems to be a worthwhile preventive measure.

Bag Development

Normally, the udder of the ewe will begin to enlarge and the teats become more erect as lambing approaches. Some ewes will "bag up" weeks before lambing, whereas others will not do so until just a few days prior to lambing time. This characteristic varies with the individual ewe, of course, but it may also tell you something about the unborn lamb or lambs. The hormonal signals that trigger bag enlargement come from the lamb. Because of this, it may be that the ewe who has a good milk supply does so partly because the ram that bred her passed on genes to the lamb that made it send a big dose of bag-promoting hormones to the ewe. Scientists studying dairy cow records discovered that a dairy farmer can increase milk production from the cows by using bulls who sire calves that cause big bag development. There is no reason sheep raisers should not do the same.

Ordinarily, the ewe will have no problems associated with the enlargement of her bag, but this is not always so. If a heavy milker gets too engorged prior to lambing, milk her out. Save the colostrum because she may not have enough of it when the time for lambing comes. Mastitis-prone ewes can be fed Aureomycin crumbles with their feed at the rate recommended on the bag for a week before and a week after lambing.

Some ewes who have had mastitis previously will have their bag enlarge to an alarming size before lambing. One year a ewe who had lost one side of her bag the year before had her bag swell so much that it dragged on the ground and was cold to the touch. We were just getting ready to sacrifice her and save the lambs when she went into labor and had beautiful twins. Her bag quickly went down, and she raised one of the lambs herself.

Generally, bagging up is just a normal sign of impending lambing. Just before lambing, as much as thirty-six hours before, the bag may be hot to the touch. This is also normal, and is one of the clues that lambs are on the way.

EVALUATION

Make a list of the ewes that prolapsed, had milk fever, ketosis, or acidosis. Any ewes that are repeat offenders for these problems should be considered as prime candidates for the truck. Except for prolapses, you should consider whether the shepherd was at least partly to blame. Either change the sheep or the shepherd.

6

LAMBING

Lambing is the culmination of both the shepherd's and the ewe's cycle. The ewe has survived countless challenges in the form of disease-causing organisms, inclement weather, competition for feed, and just plain staying alive to carry out her function as the maker of offspring, the prime mover in the perpetuation of the species. Factors such as the length of daytime periods, seasonal changes, and daily changes have interacted in complex ways with internal mechanisms to guide the sequence of physiological changes that have brought her and her unborn lambs to this point. Her instincts and her memory of previous years have served her well, and she is prepared for pushing her youngsters out into the world.

Ideally, the shepherd has worked along with the ewe, providing feed in the right amounts and of the appropriate composition at the correct times. Diseases were staved off by preventive care, and sickness was treated properly. For the shepherd each year is a new period of learning and experiment—always seeking the elusive perfect plan for care and management. Every year is an exploration of the unknown, because nothing is ever exactly the same. The skillful shepherd learns to read the signs in the flock and make appropriate changes. The shepherd has spent a lot of time watching the sheep and learning from them.

Lambing is a time when sheep and shepherd are most closely associated. Even with range flocks in the West, ewes are commonly brought to sheds or fenced lots for lambing so the herders and owners can be with the band to provide at least minimal help as needed. With the farm flock some overdevoted shepherds may literally move into the barn with the sheep to be sure to be there in time of need. Some small holders do the reverse and bring ewes into the house. Many producers let the ewes lamb in a clean pasture, just checking once a day, especially if they have culled ewes who have lambing problems in an attempt to build a low-labor-input flock. No matter what the environment, ewes and people form a close bond at lambing time that is unparalleled in the relations between man and other animals. It is a relationship that is not far different from that between husband and wife at the birth of a baby. It is an event that measures the sheep raiser against an unwritten standard and rewards those who give of themselves.

Yet some sheep raisers almost totally ignore lambing. They don't want ewes who give twins because they are too much trouble. They want all lambs born by magic, preferably in the daytime on a warm day, and they will tell you that a ewe who doesn't do everything according to their preconceived notion is stupid. These are the same people who think that a sick sheep is a dead sheep, that lambs have no will to live, that worming is a waste of time and money, that sheep can certainly winter over on a diet of corn stubble, and that their grandfather sure didn't have to do all this fancy stuff to raise sheep. One can only hope that the economic losses they will incur are sufficient punishment for these ignoramuses.

EARLY SIGNS

During the last part of gestation the pregnant ewe may have bagged up. Apart from her great size, this is the first indication of events to come. The enlargement of the udder is one of many changes, some visible, some not, that are triggered by hormonal signals from the unborn lamb. The ewe responds to the chemical signals from the lamb with chemical signals of her own, a sequence that can be monitored by measurement of hormone levels in the blood if fancy analytical equipment is available. The ewe requires no fancy laboratory facilities to tell her that events are getting ready to take place. To male shepherds all of this must be taken on faith, but I have been assured by women that early labor can be sensed by the female long before anything overt happens.

The ewe may stand apart from the flock, looking rather distant and distracted, and lose her appetite. Ewes that are getting ready to lamb for the first time will act puzzled and even distressed by the strange and unfamiliar signals they are getting from their bodies. Older ewes remember the signals, and may begin to act motherly days before their due date, often trying valiantly to steal lambs from ewes who have already lambed. The ones that are due may have a swollen, pink vulva and may lose a mucous plug and show a slight, clear, viscous discharge from the vulva.

Still further on toward the great event, the ewe's uterus will sag to a lower position. This event is easily visible because of a change in the ewe's body shape. The plumpness at hip level shifts to a lower level, leaving a hollow on each side below the spine and just forward of the pelvis. The ewe is said to have sunk. This sinking is relative to how she looked earlier, of course. Some older ewes look pretty sunken and saggy year-round. If sinking is evident, lambing will probably occur within hours rather than days.

About the same time that the sinking comes on, the ewe may seek a place isolated from the flock and dig with her front feet to make a nest. Once again, the experienced ewes and first-time mothers may react differently. Most old timers will seek shelter in the barn or shed and dig their nests in available bedding. The new mothers are more likely to dig outdoors, well away from the flock, and may not even dig at all. If a ewe starts digging, or a yearling goes off by herself, the shepherd can expect lambing in the hours that follow.

GETTING CLOSE

Some ewes will show all the early signs of lambing; some will show none at all. Don't worry if you spot none of the "official" early signs of lambing, especially if you are new to sheep raising. A sign that is obvious to an experienced shepherd may be altogether invisible to the eye of a novice. Keep looking; you'll learn.

When lambing is really getting under way, most ewes will lie down in their crude nests and show signs of strong contractions. After a push or two they will commonly get up, move around a bit or turn in place, and then lie down and try again. Sometimes they will choose a new spot if things don't go the way they want them to in the first one. When they are really in hard labor they will pull their upper lip back as they push hard, sometimes making grunting noises along with the pushing and straining.

As contractions and pushing continue, many ewes begin to make licking motions in instinctive response to their hormonal messages. Older mothers may try to claim other lambs at this point, but then they usually get back to business without need for encouragement.

Novice mothers and even some older ewes may bypass all these symptoms and just stand around until they start to lamb. Whether or not signs of lambing have preceded it, at some point the lamb and the enclosing water bag (the amnion) will leave the uterus. In a normal birth, the muscles of the cervix, the opening connecting the uterus and the vagina, will relax and the cervix will dilate to accommodate the passage of the lamb and the amnion with its contained fluid. At the same time, the pelvic bones should have separated and the pelvic opening enlarged to permit passage of the lamb and water bag without undue difficulty. The water bag sometimes breaks before the lamb has emerged, or it can protrude ahead of the lamb. When the amnion breaks, fluid pours out; the ewe has "lost her water," in common parlance.

The usual and normal presentation of the lamb is front feet first, followed by the nose resting between the tiny hooves. If this is the case, many shepherds will break the water bag to release the fluid because normal presentation generally means a birth with minimum problems, and one does not ever want the lamb to inhale amniotic fluids. If the birth were to be abrupt, as sometimes happens, the umbilical cord can rupture, signaling the lamb to take a breath, and it can inhale a usually fatal sample of the amniotic fluid.

BIRTH

In the hoped-for normal case, the ewe will push during contractions, and the lamb's hooves will show, followed by the nose, and then the rest of the body. This can take a few seconds or hours. As long as the amnion is not broken, there is no hurry. The lamb is receiving oxygen and nutrients through the umbilical cord, and it is safely cushioned by the amniotic fluid. The shepherd

Birth of a lamb: the water bag has broken (the remains are dangling), and the tips of two front hooves are visible.

should resist the urge to help when help is not really needed. As long as the ewe is not tired, and she pushes along with her contractions, give her the chance to complete the job all by herself. If the birth is too easy, either by itself or because of premature assistance by a well-meaning shepherd, the ewe may not receive enough feedback signals from helping; she may not realize that she has lambed at all, and she may reject the lamb as not being her own.

If a ewe is slow to start lambing, and the water bag breaks, or she begins to tire and reasonable progress is not being made, some assistance is in order.

There are two common causes for slowness of a normal-presentation birth. Probably the commonest is the vulva not stretching enough to clear the eyebrows of a large lamb. The shepherd can usually ameliorate this condition by running a finger firmly over the back of the eyebrows from outside of the taut vulva. Do this a few times, rather like working a tight, heavy sock over a heel; the head will generally pop out, and the rest is easy. If that fails, use your forefinger to spread some lubricant over the lamb's head and around the inside of the vulva, and try again. At the same time, pull the legs downward toward the

This birth is a normal presentation, with both hooves and the tip of the nose protruding from the tightly stretched vulva. At this point, if the water bag had not already broken, the shepherd would break it and clear the lamb's nose.

ewe's heels, wiggling them a bit as you pull. If the head is the problem, your job is to free it, so don't get overly enthusiastic about the legs.

The other common problem with a normal presentation is that the lamb's shoulders are too big to slip through the pelvis or vulva. The technique here is to pull on one leg only so as to tilt the lamb's shoulders and get one through ahead of the other. Take a firm grip on one leg, and pull it downward while you move it back and forth to help work the leading shoulder ahead of the trailing one. The shoulders will finally release with a mighty heave, and the rest of the delivery should be easy.

Once the lamb is out, the ewe may get up immediately and turn to lick it to clean it up, making characteristic gargling baas to the lamb. Let the umbilical cord break on its own. Once it is broken it should be promptly dipped in iodine. If a piece of the cord longer than a few inches is attached to the lamb, pinch the cord near the lamb's body, and pull at it with a stripping motion with the other hand to break it off closer to the abdomen, and then dip the

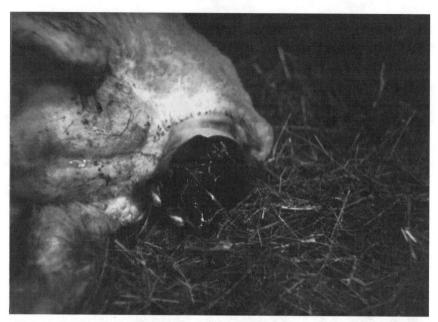

After some pushing by the ewe, the head pops free.

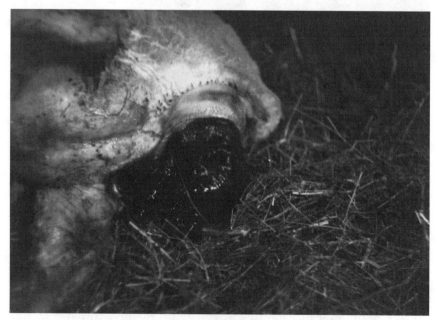

Following more contractions, the ewe has pushed out the shoulders of the lamb.

The lamb is born, still partly wrapped in the remains of the water bag, as the ewe begins to clean up the newborn. This lamb was the first of twins.

cord in the iodine. A convenient way to ensure a thorough iodine treatment is to keep the iodine in a small, wide-mouthed jar. Catch the free end of the cord in the jar and then place the mouth of the jar against the lamb's abdomen, and tip it so that the whole cord and the area around it are bathed in the iodine, the jar being pressed snugly against the abdomen to prevent spilling. With a vigorous lamb in warm weather, that is all that is required of the shepherd.

Needless to say, not all lambs have read this book, so they naturally don't do everything just as I've said they will. One thing some of them don't do is breathe. Now and then one will come out and just lie inert on the floor. You may be able to see or feel a heartbeat, but there is no breathing. There are various methods recommended by different sources to induce breathing. These include bizarre techniques such as plunging the lamb into ice-cold water to make it inhale at the shock, or dribbling some whiskey down its throat to make it cough and start to breathe. It seems to me that anyone who recom-

First meal: since that teat is a long reach, the newborn Lincoln ram must stretch for its first colostrum milk.

With a little nourishment, the lamb is up on all fours, tail flying, for a serious meal.

Two handsome ram lambs out of a champion Lincoln ewe—what more can a shepherd ask?

mends this should stay in the house with the whiskey and ice water, and send someone else to the barn for lambing.

In most cases all one has to do to induce breathing is shake the lamb a little or lift it an inch or two off the bedding and drop it. Almost any movement, including the ewe's licking and nudging, seems to help. If shaking or dropping don't work, grab a piece of straw from the bedding and use it to tickle the inside of the lamb's nose. All of these actions should take you only a few seconds.

If the lamb still won't breathe, you should promptly swing the lamb through the air, as recommended by the English veterinarian Eddy Straiton in his book *The TV Vet Sheep Book* (Ipswich, England: Farming Press Ltd., 1972), where he is pictured with a lamb flying over his head, gripped by a single, muscular hand. We prefer two hands for those slippery little ones, but the technique is otherwise the same. Grasp the lamb by the rear legs, up by the hocks, and swing it in a full circle, up over your head, and back down again a couple of times. If it doesn't breathe, repeat the procedure. The only time we have ever had to do anything else was with a head-first lamb that had a very swollen tongue blocking the trachea. I cupped my hand over its mouth and nose, and blew gently into my hand to inflate the lungs, then Teresa swung the lamb. It started breathing—as all lambs will as far as I know, if there is any spark of life in them at all.

Be careful not to hit the hapless lamb's head against the barn or the ground. If you are indoors with a low ceiling, the lamb can be swung in a circle with a high side and a low side to do almost the same job. The reason that this method works so well is that the weight of the gut falls away from the diaphragm at the top of the swing, causing an inhaling. The gut then presses against the diaphragm at the bottom of the swing—both from the force of gravity and from the centrifugal force of the rotation; this pushes air out of the lungs. The swinging also helps clear the lamb's mouth and nostrils of fluids. You should not hesitate to swing the lamb right away if the dropping and shaking fail to work, because the lamb is deprived of oxygen and could suffer brain damage quite quickly. Even a lamb with a full deck of cards is no genius, so don't make things worse.

An effective alternative to swinging the lamb is to use a chemical called Dopram V (Doxapram Hydrochloride), which stimulates breathing. All one needs to do is place a few drops, up to 1/4 cc, under the back of the tongue. It can be injected with a short needle, but it will be absorbed by the tongue even if not injected. Be sure the airway is unobstructed, because Dopram V really works, and fast, too. This is a drug that must be obtained from your veterinarian. Some vets are reluctant to give Dopram to clients, but used carefully it is safe and very effective. I always carry a small bottle of it and a small syringe in my coat pocket during lambing. Be careful not to get it on your skin, because it will affect your breathing too.

Not all lambs arrive in the so-called normal birth position with both front hooves forward and the head nestled in between, so be prepared to deal with other possibilities. Some of the various birth positions present few difficulties, but others are a bit tougher to deal with. I'll describe the various possibilities and suggest solutions.

One Leg Back

If only one leg and the nose are showing, don't try to get the other leg turned forward and out. Roll the ewe on her side so that she lies on the same side as the turned-back leg. Thus the ewe should lie on her right side if it is the

Swinging a slow-to-go newborn lamb like this will start it breathing almost every time.

lamb's right leg that is turned back. In other words, roll the ewe on her side so that the turned-back leg is up. Grasp the leg that is out and pull as you work the vulva over the lamb's head. At this point, the lamb will usually either pull free or the trailing shoulder will hang up on the ewe's pelvis. In such a case you want to pull on the free leg and twist the lamb at the same time. To

153

do this, pull on the leg as you move it toward the ewe's spine. At the same time, pull on the head as you move it toward the ewe's hocks. These motions will rotate the lamb and sort of unscrew it from the pelvis. This may take quite a few tries and quite a bit of force, but keep trying. The shoulder will finally pop free, sometimes very abruptly, landing you on your rear end, and the rest of the delivery should be easy. With this and other assisted deliveries, you can give the ewe a chance to finish the job once you have solved the problem, but don't give her too long because the lamb has already been through a lot, and should be gotten out promptly either by the ewe or the shepherd.

Head Only

If only a head is showing, the ewe is unlikely to be able to deliver on her own. You should feel with a finger to find out if the lamb's legs are both doubled back or if one or both of them is next to the side of the head. If a hoof is next to the head, try to pinch it with your fingers and pull it out. Those little hooves are mighty slippery and hard to hold onto, so you might try a loop of twine or umbilical tape to snare it. Do the same with the other hoof if it is there. If both hooves can be brought out, the birth should proceed normally. If the lamb's tongue is swollen and bluish in color, go ahead and pull the lamb promptly. If you can only catch one hoof, proceed as with one leg back.

If the head is showing and both legs are trailing back, you will have to get one leg out. Push the head back through the pelvis, making sure that it doesn't turn back. Feel at the shoulder to find a leg. Start at the shoulder, using one or two fingers, and straighten the leg and bring the foot through the pelvis. Now bring the head back through, and bring both head and the one leg out, and proceed as with one leg back.

If the head has been out or in the birth canal and is badly swollen, you may not be able to push it back inside. In this event, use lots of lubricant and try to work a hand or even a finger in behind the head. Try to grasp behind the head or on a shoulder. Get hold of anything you can. Patience is the key here, and strong fingers don't hurt a bit either. Pull and tease and twist and rock the lamb back and forth. Usually the lamb can be saved with enough slow and persistent work. All too many people give up and cut off the lamb's head or take some other drastic solution without making a fair try to work the lamb out.

Extra-Large Lamb

Even a normal presentation can be difficult if the lamb is very large. It helps in these cases to have an assistant hold the ewe upside down on her spine.

This is easier said than done. The best way is to tie a twine between the back hooves of the ewe and loop it over the assistant's neck and shoulders. The assistant should be straddling the ewe, facing her posterior. One leg can be held in each hand and part of the weight carried by the twine. With the assistant holding the ewe in this position, the shepherd will have an easier time twisting and pulling the lamb out. With big lambs, try to get one leg past the pelvis ahead of the other because the shoulders may be too big to go through together.

Front Feet Only

If the front feet are out but not the head, this means the head is turned back against one side of the lamb. You will have to put the feet back inside the ewe in order to get the lamb back far enough to be able to turn the head forward. In order to keep track of the hooves, tie a piece of twine or umbilical tape to each one before pushing them back in. Then push the lamb back into the uterus, and try to bring the head around to lie between the legs. Once the head is brought around, draw it out while bringing the legs out with the help of the twine or tapes. Once the head and legs are out you are back to a normal delivery position.

If the head cannot be guided out with the hand, a loop of three-inch-wide gauze bandage, twine, umbilical tape, or insulated electric wire can be used. Twist the gauze or whatever to form a loop that fits over the lamb's head, and tie a knot if you wish. Slip the loop over the lamb's head with the loop behind the head. Let the long end of the gauze come out from under the chin. The loop and gauze are like a collar and leash on a stubborn dog that is backing away from you. Then, pull on the "leash" and on the twine attached to the hooves. This should draw the lamb out. If you use a wire noose, be very sure that the sharp end of the wire is covered so as not to cut the uterus. Also be sure that a fold of the uterus is not pinched in the loop. There are lamb pullers on the market for this purpose, but the gauze bandage or even twine is really more satisfactory in my opinion. This whole procedure is aided if the ewe is held upside down on her spine as described for the extra-large lamb.

Hind Legs First

The hind-legs-first presentation is really almost normal. If you are not sure how to tell a hind leg from a front leg, take a look at a lamb and study its legs. This will tell you more than I could even if I wrote about it for pages. Once you know you are looking at hind legs, just assist by pulling, being careful not to crush the lamb's ribs in the process because you are pulling against them. Pull gently while rocking the lamb from side to side to free the ribs from the pelvis. Once the ribs are freed, it is wise to get the lamb out quickly and clear its nostrils so that it does not breathe in any amniotic fluid or mucous.

Tail Only

If all you see is a tail, push the lamb back in and bring out one rear leg at a time by hooking your fingers around the hocks. When both legs are out, proceed as with hind legs first. The headback lamb can be handled in this way too by turning it around and delivering it backside first.

Crosswise Lamb

If the ewe strains and pushes to no avail, slip your lubricated hand and arm in and investigate the situation. If a lamb is crosswise, turn whichever end is closer toward you and proceed as with the above cases.

Closed Cervix

Another possibility when the ewe works hard but makes no progress is that the cervix is not dilated. It can help to gently manipulate and massage the cervical area with a finger, which sometimes will cause the cervix to relax and dilate. Sometimes the cervix will not dilate (ring womb). In such cases an injection of 100 ml calcium borogluconate subcute in several sites may help. In other cases the pelvis has not spread enough, and the pulling procedure will take care of that. With valuable animals a Cesarean section is the last resort.

Twins Tangled

It is hard to give any rules for tangled twins except patience. The shepherd has to feel carefully and follow the legs with his fingers to find out which legs belong to which lamb. Once a lamb is identified, it can be moved to one of the birth positions. It is a matter of pushing protruding legs back and untangling the lambs with care. A special effort must be made not to get the umbilical cords caught. Fortunately for all of us, tangled twins and triplets are more often found in books than they are in the lambing barn. Lots of exercise during gestation is the best insurance against tangles.

Failed Efforts

If any of the above techniques fails completely, the shepherd will have to decide whether to call a veterinarian. If the ewe or lambs are sufficiently valuable, a Cesarean section may be the answer. Pulling harder and harder will probably kill the ewe as well as the lambs. If you opt for the C-section, it may be a good idea to call ahead and take the ewe to the vet's office. The vet will be ready and waiting, and will have better facilities than your barn. Also, you save yourself the cost of an on-farm call.

NEWBORN LAMBS

Once a lamb is out of the ewe, is breathing, and has been treated with iodine, then what? In some cases, nothing else need be done right away, but it is usually worthwhile to take care of a few minor items. In cold weather, the ewe can use a little help in drying off the lamb. Rub it with a towel, an old sack, a handful of straw, or anything handy. If the weather is really cold, say below zero, more serious measures are in order. The ears of lambs are susceptible to freezing in zero and below-zero weather. You must get them dry, and the best way to do that is to use a hand-held hair dryer. If the ewe is doing a good job of licking there may be no reason to use the hair dryer on any part of the lamb except the ears, but do get them dry. Be careful with the hair dryer, because you can burn the lamb if you are overly enthusiastic.

If a lamb is chilled, as may be the case for one born unattended, it will have to be warmed up. It can be placed under a heat lamp or warmed with the hair dryer. For really chilled lambs, the best thing is to immerse them in warm water. Be aware that a chilled lamb may be too cold to shiver. A shivering lamb is doing so to get warm and may be doing just fine, though it might need a little drying if it is wet. To warm a chilled lamb, get a pan of water that is warm to your hands but not hot and immerse the lamb, keeping its nose above water. Let it stay there a while, soaking up heat, then remove it and get it thoroughly dry with a towel and hair dryer. There is always a chance that you will rinse off enough of the lamb's scent so that its mother will reject it, so don't scrub it, just let it soak, and let her smell it before you scrub it with a towel. If possible, you can do the immersion right in front of the ewe so she sees and smells the whole thing.

Some shepherds take a lamb into a warm room or into the house to warm up, but the water method is much quicker. Some lambs are warmed in an oven, or so I'm told. For heaven's sake, don't ever put a lamb in a microwave oven. Whatever method you choose, be sure that the lamb is warmed clear through and not just on the outside. The reason for this is that in a cold lamb the internal blood vessels are fully expanded. If the external blood vessels are opened by warming, a drop in blood pressure will occur, with resulting shock. Warm slowly and fully.

Some producers build warming boxes to use with chilled lambs in the barn. There are many designs, but basically they are boxes with warm air circulating through them. The warmth can be from light bulbs or from a hair dryer attached. Design them so the lamb cannot contact the hot light bulbs, or so the warm air from a hair dryer does not blow directly on the lamb. The temperature should be 100–120° F. If you are handy with electrical things, use a thermostat to control the temperature.

The usual reason for a cold lamb is a lack of food. A lamb that gets one good meal from the ewe will rarely get chilled. That early meal of nourishing colostrum seems to really fire up the internal furnace. Check the lamb by sticking a finger into its mouth. If it feels cold to you, the lamb is chilled. It needs food and warmth. If you have already warmed it, get some food into it. It is hard to say which to do first, because both are important.

The first food should be colostrum, that first, very thick milk that comes from the ewe. I'll even go so far as to say that the first meal *must* be colostrum. Colostrum has three main effects. First, it provides quick nourishment in just the form the newborn lamb needs. Second, it is mildly laxative, and it helps the lamb to get rid of the first tarry, sticky feces (the meconium) that it is born with. Quick energy and a laxative are both good things for a lamb, and the energy will warm the lamb quickly.

The essential thing about colostrum is that it contains antibodies from the ewe to help the lamb fight off disease until it starts making its own antibodies. Without the protection from colostrum, the newborn lamb is completely unprotected. In human beings, some antibodies are transferred through the placenta to the unborn baby, but with lambs this is not the case and they must have colostrum. Colostrum that contains a large amount of imunoglobulin-G's will often be very thick and have a high weight per volume. However, yellow color and a thick consistency are not indicators of quality. If you really want to be quantitative, buy a colostrometer from a veterinary supply house. High-quality colostrum will contain 50 mg or more of the IgGs per ml. You should reserve and freeze a quantity of the high-quality stuff to use when a ewe doesn't produce the premium grade product. There are products marketed as colostrum replacements, but they are not nearly as good as the real thing. They can be used as supplements, but not as a replacement.

Colostrum antibodies are large protein molecules that can be absorbed through the walls of the intestine of very young lambs. The ability of the intestine to absorb large molecular weight proteins is limited. If ordinary proteins are in the intestine, they will be absorbed, and the capacity of the intestine to absorb antibody proteins will be used up by the ordinary proteins. Thus, an early meal of ordinary milk or some prepared mixture will in effect take up the space that could have been used for antibody absorption. For that reason never give a lamb a feeding of milk or other fluid until it has had a meal of colostrum either from its mother or from the shepherd, unless of course there is no possibility of its getting colostrum for some reason—such as the death of the mother, the lack of milk letdown, or some other cause. You should have saved some real colostrum for bottle feeding to cope with this possibility. A lamb should ideally get colostrum within thirty minutes, and in no case later than six hours from birth. Within the first twelve hours of life, and lamb should get about 10 percent of its body weight in colostrum.

Colostrum supplements will not contain antibodies that are specific to the disease organisms on your farm. I did talk a dairy farming neighbor into letting me give one of his pregnant cows an injection of the bacterin for immunizing against *Cl. perfringens* types C and D. Then he saved most of the cow's colostrum for me to put in the freezer to use with lambs. It was very effective, although not as nourishing as sheep colostrum.

When thawing frozen colostrum or mixing colostrum substitutes with water, keep the temperature below about 125° F, or better yet just room temperature, to avoid destroying some of the IgGs.

As part of the normal routine to get the lamb started feeding, first clear the ewe's teats by stripping them with thumb and forefinger to expel the waxy plug and get the first milk flowing. Then you can let the lamb try to find the teat on its own, and a vigorous one will do so. Give it a reasonable time to do the job itself. It is generally of no use to try to put the lamb to a teat if it is not already trying to get up by itself, anyhow. Once it starts trying to get to its feet, it will be ready to suck. It can be guided to the teat with your hands by nudging it gently from the side or rear the way a ewe would do. The head can be held under the chin and put to the teat. With some lambs, just aiming them in the right direction is all they need. Don't try to grab the head and force it to the teat. The lamb will fight you and tire itself. If the lamb doesn't suck readily, avoid frustration by using a bottle. Milk about two ounces of the ewe's colostrum into a clean bottle, screw on a lamb nipple, and feed the lamb. If it doesn't want to suck on the bottle either, try covering the top of its head and eyes with the palm of your free hand; this simulates the feel of the ewe's bag and thigh. You can also try tickling the lamb under the tail to get the sucking instinct stimulated. Touching the roof of the mouth with the nipple is sometimes helpful too.

If these methods fail, and they sometimes will, you should get a meal into the stomach by tube. A lamb that will not suck on its own is probably too cold or weak or both. There are a number of ways to get a few ounces of colostrum into the stomach. I prefer an attachment for a glass-metal syringe called a Lambprobe that is made by the McGrath Company, P.O. Box 148, McCook, Nebraska 69001. The Lambprobe is a stainless steel tube with a little bulb on the end that can be guided down the throat of the lamb into its stomach. The lamb is placed on your lap, lying on its side, and the probe is inserted with the right hand as the left hand is held to the underside of the head and neck so the bulb can be felt going down. After a try or two you will find that the Lambprobe is quite easy to use. Ours attaches to a 35 cc syringe, which allows one to give a little over an ounce of colostrum per insertion. A pistol-grip syringe would be even handier. There is a similar device called a safety-ball nozzle pipe available from Wooltique (see appendix 6 for address) that attaches to a drench syringe; however, the ball may be a bit large for very small lambs.

You can make a stomach feeder from a 60 cc plastic syringe by attaching either a length of quarter-inch (inside diameter) clear plastic tubing or an ordinary rubber catheter that can be obtained from a hospital or nursing home. If you use the plastic be sure to smooth the cut edge so that the lamb's throat is not abraded or cut. Smooth it with sandpaper or by heating with a flame briefly. Both of these tubes, particularly the catheter, are flexible, so they cannot just be shoved down the throat as with the metal tubes. Let the lamb swallow the tube as you advance it.

There are other arrangements for stomach feeding that I do not consider quite as handy. You can attach a funnel to one end of plastic tubing. Or, you can use an enema kit purchased at a hospital, drugstore, or nursing home. It has a plastic bag with a long tube attached and a clamp on the tube to serve as a valve. The bag can be hung on a nail, which frees both hands to insert the tube. There are also rubber bulbs with tubes attached, sold by suppliers.

When you stomach feed a lamb, be sure not to give it too much colostrum. Most authorities recommend feeding a lamb less than a week old only one to

Feeding a lamb with a Lambprobe is simplicity itself. Fingers under the throat, holding the neck straight, can feel the probe go down.

two ounces of colostrum every two hours for the first four feedings, then changing over to two to four ounces six times a day. Err on the side of less rather than more, because overfeeding will give a lamb diarrhea (scours) from which it may not recover. Remember—no matter how many gadgets you have for stomach feeding—that you are just trying to get the lamb started so it will suck on its own mother or on a self-feeder. When a lamb sucks, the milk passes around the rumen to the true stomach where it belongs.

A really weak lamb can be given nutrition in the form of a solution of glucose that is injected. There are two concentrations of the solution on the market (5 percent and 50 percent). There also are dextrose solutions. It doesn't matter which sugar is used. Most people prefer to use the 5 percent solution because it does not invite dehydration. A total of about 2 to 3 cc per pound can be given subcutaneously at several sites (no one site should receive more than 10 cc). There are also solutions that contain amino acids and electrolytes, but the plain sugars do the job. Repeat every four to six hours if needed.

You can also inject the whole amount into the body cavity. Veterinarian David Henderson (see his book in appendix 6) suggests mixing 40–50

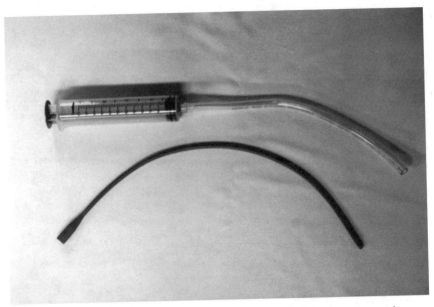

A plastic syringe with either a length of plastic tubing attached (as shown) or with a rubber catheter (below) can be used to put a meal in a lamb's stomach.

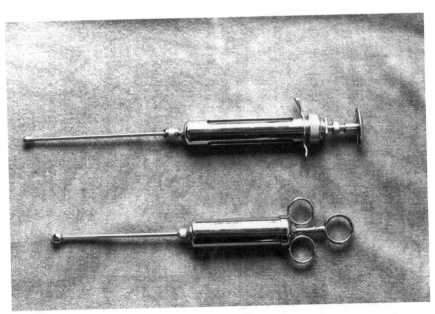

Two types of stomach feeding devices: top, a Lambprobe-equipped glass and metal syringe; bottom, a safety-ball nozzle pipe on a drench syringe.

percent glucose or dextrose solution with an equal amount of freshly boiled water. You want to give about 5 ml of the solution for each pound of weight of the lamb. So, maybe 50 ml for a big single to 30 ml for a small triplet. Use a large syringe—like 50 ml or so—and, using a needle, draw half the amount you need from the concentrated sugar solution. Remove the syringe from the needle and draw in an equal amount of the hot water. Place a one-inch needle on the syringe and draw in enough air that you can mix the solution a bit. Then, needle up, force out the excess air. Standing, hold the lamb by its front legs, hanging down against and between your thighs. Insert the needle downwards at about a 45-degree angle and an inch below the navel and a half inch to either side of the navel, and inject the contents slowly. The needle should slip in with no resistance. If you feel that you are hitting something, withdraw and try again. You don't want to inject into the bowel or intestine.

While you are shepherding one lamb, the ewe may be getting ready to have another one. If she has been jumpy and nervous and has not let the first lamb nurse, it may be that she feels the signs of another one coming. Be patient in getting the first lamb to the teat if another is on the way. A ewe who is going to have another lamb usually holds her tail straight out or even somewhat elevated. Also, if a water bag containing a yellow fluid protrudes from her vulva after the birth of one lamb, probably another will arrive.

A water bag filled with red, bloody fluid usually, though not always, means that lambing is over. The sure sign that lambing is over is the appearance on the water bag (either filled with fluid or empty) hanging from the ewe's vulva of a white, wormlike piece of tissue, either attached to it or corkscrewed around it.

Once a ewe has had her last lamb she will generally get down to the business of being a mother and will lick the lambs clean and help them to her bag with nudges from her nose. Some ewes may show a preference for one lamb over another. This usually passes with time, but be alert for a ewe that rejects one of a pair, because the ignored one can starve if not allowed to suckle. Some ewes will not accept their lambs at all, especially first-time lambers. In such cases, the ewe has to be confined in some way to allow the lambs to nurse her. I'll discuss methods for this under Handling.

With most ewes and their lambs, all that is needed is to pen them up in a jug (pen) where they can get fully acquainted. Don't let them run with the flock if you can help it, because the lambs and ewe can get separated. Also,

The white, wormlike structure (next to the finger) on this water bag filled with red-tinged fluid tells the shepherd this ewe is not going to have any more lambs, at least not this year.

other ewes may injure the newborns, who will try to nurse on any teat that is handy until they learn to recognize their mother, and most ewes get pretty irate when a strange lamb tries to suck on them. So, pick up the lambs by their front hooves and drag them slowly across the bedding to the pen. Dragging leaves a trail of scent for the ewe to follow. If she returns to the birthing place, as many of them will, try again until you can close her in the pen. Stand back and admire the newborns with their mother for a moment and then get back to work. If you have used a heat lamp, consider removing it. The extra heat seems to encourage pneumonia if used too much, and the lamps can start fires.

NUTRITION

The ewe needs to have her food intake adjusted for a few days to help her make the transition from lamb maker to milk factory. The ewe's milk-secreting organs have already been stimulated by the hormones from the unborn lamb, and they are ready to go into production. But full production is not really what the shepherd wants right at first because the demands of the lambs may not quite be up to the ability of the ewe to make milk, and an overinflated udder is an invitation to bag problems or reduction in milk output.

In order to keep the milk supply at a lowered level for a few days, no grain should be fed. The ewe will relish a big tab of good hay right after lambing, and she can be given about four pounds of hay a day without creating any problems. Once the lambs are nursing strongly, she can be put onto a full lactation ration by increasing the grain in steps. You can judge when to bring grain back into the ration by feeling the ewe's bag. If she is being milked out, her bag will be soft and pliable even though it may be large. Be sure to check both sides of the bag, because sometimes a lamb or lambs will nurse only on one side or more on one side than the other. When this happens the shepherd can try to introduce the lamb to the other side, but it is hard to change their habits once they are established. If a ewe is too full, milk her out and save that colostrum for future use.

Some shepherds give their ewes a bucket of warm water with some molasses in it for the first after-lambing drink. There is nothing wrong with this, and most ewes have enough of a sweet tooth to enjoy it, but it probably isn't really necessary. An exhausted ewe might benefit from the energy and warmth, though. Most ewes are perfectly happy with plain cold water, however.

In the foregoing discussion, I have assumed that the ewe has a full bag that the shepherd is trying to keep under control. Some ewes will be slow to come into their milk, however, and may need a little encouragement. For these you can keep feed at late-gestation levels but improve the quality of the hay they get. You should encourage slow milkers to drink water by giving them salt, and even give them warm water if that is practical for you to do. Don't forget to check the lambs of ewes who are not yet milking to capacity. A hungry lamb will usually complain loudly, but the shepherd should pick it up anyway. A hungry lamb feels light, thin, and hollow, and it may be cold as well. It will not have the round, full belly of the well-fed ones. Be sure to check the lambs, because they can die of starvation even in the pen with their mother.

The ewes who started off with little milk may come into full capacity suddenly, and this needs to be controlled. Check bags daily, or oftener, and make adjustments in feed as needed. There is no substitute for experience, so get in the pens and feel the bags frequently. You'll learn fast. If you run into a problem you don't know how to solve, ask a dairy farmer or someone who milks goats. They'll be able to give you sound advice.

With large operations in which hand palpation of bags is not possible, Aureomycin can be fed at recommended rates in the ewe's grain ration to reduce incidence of mastitis. Mastitis rarely shows up in the first day or two after lambing, except in ewes who had it in previous years. It is worthwhile to check, anyhow. Milk a squirt of milk into a shallow, dark-colored container or even into the palm of your hand. If it is stringy, lumpy, watery, or bloody, the ewe may have mastitis and should be treated as discussed in Lactation.

ENVIRONMENT

If the ewes lamb in an insulated barn or in an open shed, there is a need for small pens for the ewe and her newborns. These pens are called gaols (jails) by the British and jugs by Americans. The size of lambing jugs has undergone negative inflation in the past couple of decades. Authorities used to recommend jugs as large as six by eight feet. The trend today is to a jug of four by four feet, or even three by four. We use four-by-four jugs and find them to be plenty roomy, and some of our ewes are quite large. Small jugs are a benefit, because one can cram more of them into a given space if barn area is limited.

There are many designs for jug panels. Most lambing barns have a mixture of types that reflect the changing ideas of the shepherd. Also, a lot of equipment such as panels is bought and traded at auction sales and the like, so some panels are home built while others were bought secondhand. Apart from details of construction, the main choice in design is between solid panels and ones that the ewes can see through. I personally favor solid panels—all else being equal—because the ewes cannot see through them. Some ewes are highly protective of their offspring and will try to fend off neighboring ewes by butting through the panels. If they can't see the other sheep, they won't butt. I'm not the least bit concerned about them hurting their hard skulls, but they can reduce a panel to rubble in short order, and then I have to repair it.

Our panel collection is as motley as any, including specimens as bizarre as board panels made from wood stripped from vintage railroad boxcars, but my favorites are made of a solid sheet of plywood. A standard 4-by-8-foot sheet of plywood can be sawed into three panels that are 48 by 32 inches with zero waste. Because of the high humidity of most barn floors, I use plywood with exterior glue. You may be tempted to use half-inch plywood, but five-eighths is a lot sturdier and is a better long-term investment.

Panels can be linked together with hinges or elaborate locking pins, which is perfectly fine if your budget includes money for the hardware. My attempts at fancy designs with metal hooks and keyways were all flops in actual practice, and we assemble all of our jugs with old baler twine, a product that is always in surplus supply in a lambing barn.

I drill large holes in the sides of the plywood panels, two on each short side, and tie the twines through the holes. With the cheaper grades of plywood, some of the holes will break out, a problem that can be solved by getting a better quality of plywood, or by nailing a one-by-four of lumber down the short sides as strengthening.

A double row of jugs can be made in an open area by using a steel-mesh hog panel as a divider and tying four jugs on each side to give a free-standing group of eight. Active ewes may deform the jugs from squares to parallelograms with this arrangement, a minor nuisance that can be corrected by moving them back to a square each day or by putting a diagonal brace across one of the jugs in each row. The ewes can see one another through the hog panel, of course, but the steel will resist their onslaughts until they tire of the game.

Feeding in the jugs is managed with simple wooden feeders that hang on the sides. Some greedy ewes will eat their neighbor's hay as well as their own with such an arrangement if they can reach it, and there is something to be said for feeding a tab of hay on the floor in a corner to stop the stealing by larcenous associates. One advantage of the solid plywood panels is the absence of crosspieces that the ewes might stand on to extend their reach into adjacent pens. Grain can be fed in the same feeders or in small buckets. Water is also usually given in buckets, although that can add up to a lot of labor.

After the lambs and the ewe have been together in the jug for a few days, they should be moved together with some other ewes and their lambs to a larger community pen. This setting will give the lambs a chance to learn to distinguish their mother from other ewes. The ewes do a brisk job of training by attacking lambs other than their own who try to nurse them or even approach them.

Mother and lambs soon learn to recognize one another by sound, sight, and smell. After a few days in a community jug, the whole bunch can be turned out to join the flock.

Some ewes are so protective of their lambs that they will injure other lambs, sometimes seriously. We had some ewes who broke legs and even killed strange lambs in an excess of motherly zeal. I cured them of this habit by treating them like sheep treat one another, that is by whacking them on the hard part of their skull with a board every time they tried to attack a lamb. After a few firm whacks, they left the other lambs alone. They were never quite bright enough to figure out that it was a board and not a lamb that was hitting them.

Even without nasty ewes, community pens are always somewhat of a three-ring circus, and it is helpful to have a place where the lambs can get away from the ruckus to sleep. All you need is what is called a creep area that is fenced off from the main pen by a panel with slots wide enough for the lambs to squeeze through but too narrow for the ewes. The use of a creep area for rest and relaxation at this early time in the lambs' lives is also good training because when you start to offer them solid food it will be in a similar setting. Even with newborns, it is good training to have some leafy hay and a little freshly ground grain mixture available in the creep area for them to experiment with.

All of the quarters should be well ventilated but not drafty. There are many variations on barn design, but a basic principle is to try to move air through to sweep away excess moisture and gases, mostly ammonia, from decomposing bedding. In cold climates, water or frost will condense on uninsulated ceilings and create a nuisance. The requirements of a barn are so much a function of local conditions that I don't intend to discuss barn design except to suggest that you visit other sheep raisers in your area, see what they do, and determine how they like their setup. Keep your eyes and ears open, and make your own judgments, because it is a human trait to defend one's own property or design even if it is deficient in some ways.

If you lamb in warm weather and choose to lamb in a pasture or lot, then maybe you don't want to bother with jugs. It is your choice. It is best to keep the ewes with lambs separated from those yet to lamb—the drop band—so move the ones with new lambs to a second lot or pasture.

With barn or shed lambing, when you turn ewes and lambs out of the community pens and into the flock, also keep the ewes with lambs separate from the expectant mothers in the drop band.

The ewes who are ready to lamb need peace and quiet, an environment difficult to maintain when lambs are around. In addition, the lambs will try to suck on the ewes who haven't lambed yet, and many times the ewes will let them get away with it. This can deplete the ewe's supply of colostrum to the point where she may have none left for her own lambs when they are born.

HANDLING

At lambing time most ewes become calm and more trusting of a familiar shepherd. This greatly facilitates giving assistance if it is needed. Some of our ewes even seem to welcome a leg against their shoulder for them to push against when they are trying to birth their lamb, and they rarely show fear of humans during actual lambing. Others are wild of course, but most do calm down relative to their usual behavior.

First-Time Mothers

The major exception to this lambing time tranquillity is with first-time mothers, especially those who were bred as lambs. The signals that they get from the lamb's and their own hormones seem to confuse them, and they react by getting spookier than ever. It is a great stroke of luck if first-timers are in the lambing barn, where they can be confined when their time comes without their undergoing undue excitement. A large majority of them will lamb wherever they happen to be when the time comes, unlike older ewes who like being in the barn for the event. The wild ones seldom choose a place to the liking of the shepherd, but such is life. If the lambing is routine and no assistance is needed, you just wait until she lambs, then move in and take care of any needs. A flighty first-time ewe may run away in a panic if disturbed, leaving her lamb behind. If so, back away and let her come back to the lamb. In inclement weather you may have to get the lamb and ewe into the barn. Pick up the lamb by its front legs, and walk slowly away with it, letting the ewe see it and smell it as you go so she knows it is her lamb. As often as not the ewe will return to the spot where she lambed and sniff the ground, seeking her lamb. When this happens, start over again to try to tempt her to the barn. It helps to have a dog or another person shoo her toward you and the lamb. Each situation is unique, and some will really challenge your ingenuity. One can just give thanks for the occasional first-time ewe that behaves ideally and follows quietly into the barn without a lot of fuss and theatrics.

To reduce the problem of having a ewe lamb in a place that is not particularly convenient for the shepherd, it is useful to keep the drop band in a yard right next to the barn so that lambs are not born in a remote corner of a pasture in the middle of the night where they are not noticed. The flock will soon learn to come in every night once they are used to the daily move. If they are reluctant to move at first, go out at night and approach them from the darkness as you talk to them or sing to them, all the time moving them toward a lighted yard or barn. Their night vision is not good, so they will move toward the light readily rather than rush off into the darkness. After a few nights of this they will move to the desired area as soon as they see you coming to herd them. Cold weather is your ally in this if the ewes were all sheared in late gestation, because they will go to the barn at night to keep warm.

Orphan Lambs

Lambs may become orphans—or bummers, as they are sometimes called—for a variety of reasons. Some ewes die in lambing, leaving bummers. Other lambs have to be taken from their mother because she cannot nurse them, as is the case when one lamb is pulled from a set of triplets. It is an unusual ewe who can provide full fare for triplets, though many can. For one reason or another there are usually a few extra lambs to take care of, and the best thing to do with a bum lamb is to graft it onto another ewe, that is, get her to accept it as her own lamb. I'll discuss raising lambs on artificial milk or other substitutes for ewes' milk further on.

Grafting Lambs

There are many ways to graft lambs, most of which fail because a ewe can easily identify her own lambs by smell, and she will promptly reject a stranger that the shepherd tries to foist off on her. We did have one ewe who would accept any lamb with alacrity regardless of its size, odor, or color. Most ewes, however, are not that agreeable and must be tricked into cooperation.

One hears of recommendations for forcing a ewe to accept a lamb by restraining her. The idea is to confine her in such a way that the lamb can get to the bag to nurse, but the ewe cannot step away or turn to hit the lamb with her nose or head. Such stanchion arrangements will allow the lamb to get some nutrition, but many ewes never accept the lamb no matter how long they are kept stanchioned. In my experience, stanchioning has limited application, though many producers swear by it.

Rather than using force, I prefer trickery for grafting. Almost every sheep raiser will tell you a "foolproof" way to graft, though of course no method can be totally guaranteed. One traditional technique is to graft a lamb to a ewe whose lamb has died by skinning the dead lamb and placing its skin over the graftee. The skin can be cut along the midline of the abdomen and the legs skinned out by pulling the skin off like a sock. Then the lamb to be grafted can have its legs slipped through the empty legs or else slits can be cut. This method seems to work most of the time.

There are also chemical approaches that depend on confounding the ewe's sense of smell or disguising the smell of the lamb. There are commercial preparations on the market for this purpose such as Franklin's Mother-Up. Odorous substances such as menthol creams, perfume, and pine tar all have their adherents. Although this method is widely talked about, I don't personally know anyone who has found it to be successful. A ewe's sense of smell is too good to be easily fooled or inactivated.

As far as I am concerned, there really is one foolproof method and I hope it will work for you. This method has two critical factors: first, the lamb to be grafted must be given to the foster mother before her own lamb is born and, second, the graftee must act like a newborn, be wet, and smell like a newborn.

If you are prepared, meeting the first qualification should present no big problems. Always keep in mind which lambs need to be grafted so you can act when the time comes. The orphan should be all ready to give to the ewe when her own lamb is almost born, in other words, hanging out and being pushed by the ewe.

To meet the second requirement, first tie three of the graftee's legs together with baling twine (another use for twine!) so that it cannot get up and run around and attack the teat like the veteran it is. In addition, the lamb should be completely wet, all over. It is much to be preferred if you can soak it with the ewe's own amniotic fluid. Have a small bucket ready, and pop the bag hanging out of the ewe into the bucket to catch the fluid. You can then sit the orphan in the bucket, tail-end first, and wet it all over with the birth fluids. If you did not catch any of the ewe's fluids, soak the orphan in warm water. Just before you present the graftee to the ewe, you can rub it over the exposed part of her own lamb to get a little odor on it. Using warm tap water works almost every time, and the use of the ewe's own fluids should give you a perfect record.

Ironically, what usually happens with this method is that the ewe has twins and you have to take the lamb you just successfully grafted away to try on another ewe. One year we grafted a little buck lamb three times, and each time had to take him back because the ewe had twins. We finally got the poor little fellow a permanent mother on the fourth try, even though he was practically shopworn by then. You can take the twine off the adopted lamb's legs after the ewe's natural lamb is up and walking actively.

The tied legs and wet lamb trick will sometimes work after the ewe has had her own lamb, but the odds of success are much lower. Once she has smelled her own lamb, she is pretty suspicious of others. To aid in getting a ewe to take orphans after the fact, it sometimes helps to put a short panel with slots in it in one corner of the jug. That way the lamb can retreat to the corner to get away from her constantly checking its smell and hitting at it. Generally, the orphan and the natural lamb will sleep cuddled together in the protected corner so that their smells gradually mingle, and the ewe may eventually give up and accept them as a pair. I think that this method is much better than stanchioning the ewe, because the stanchion makes the ewe nervous and upset, which makes matters worse. With a hideaway, the orphans can escape from an aggressive ewe, and then can sneak out and get a meal when the ewe is otherwise occupied or distracted.

If all of this sounds like too much trouble to you, and you figure that you might as well just give the lamb milk replacer, let me assure you that it is worth the effort. A lamb will grow better and be healthier with real sheep milk and a mother's loving care. Not only that, the cost of ewe's milk is a fraction of the cost of replacer. You'll feed that ewe about five pounds of hay and two pounds of corn a day at a cost of about twenty-five cents. If you feed the bum a lamb replacer, it'll drink about two quarts, and those two quarts will cost you about a dollar. Even not counting the better quality of real ewe's milk, the ewe will feed two lambs for a fraction of the cost of one on replacer.

Bottle Lambs

If you are not successful in grafting a lamb onto a ewe or if you have extras for whatever reason, you will have to raise them to weaning on some sort of fluid diet. Remember to be sure that the lamb gets several meals of colostrum the first day before any other oral nutrition is given. After that it can be raised on milk or on a replacer. We have used goat's milk with success, and many growers use cow's milk, but neither one really is a substitute for the richer sheep's milk. We tried both goat milk and replacer and found that the lambs prosper best on the replacer. The replacer of choice in the United States seems to be Land O' Lakes. I have made no comparative tests with other brands—which may be just as good or better—but the Land O' Lakes is said to be superior by many producers. Whatever brand you use, don't switch brands or even switch lot numbers of the same brand if you can help it. Lambs are very sensitive to changes in their diet, and I have heard of them refusing to eat replacer made from a different batch lot of the same brand.

Whether the lamb gets real milk or replacer, it is important that the shepherd limit consumption in some way. If lambs are fed from bottles, the shepherd must control the amount the lamb gets because too much will cause incurable scours.

Here is Teresa's schedule for bottle feeding; it has been highly successful for us.

First day: 1–2 oz. colostrum every two hours for at least four feedings if possible
1–2 days: 2–3 oz. 6 times a day
3–6 days: 3–4 oz. 6 times a day
1–2 weeks: 6–8 oz. 4 times a day
3–4 weeks: 8–10 oz. 3–4 times a day
4–6 weeks: 12–16 oz. 3 times a day

This is a very conservative feeding schedule for small lambs such as triplets, quads, and up, and adjustments can be made according to the size of the lamb. Remember that it is better to underfeed than kill by overfeeding. However, a large, healthy, hungry lamb may progress through the schedule more rapidly.

A simpler way to limit feeding is to let the lamb feed as it does when nursing a ewe. If a lamb is given replacer in a self-feeder, it will eat when it is hungry and will seldom overeat. It is important that milk be available at all times so that frequent small meals become a habit. If the feeder is allowed to get empty, some lambs will gorge when it is filled. In addition, the milk should be kept cold, which not only prevents spoiling, but helps to limit the lamb's intake. To keep milk cold, ice packets or plastic jugs of ice can be floated in the milk to chill it without diluting it.

There are a number of lamb self-feeders on the market such as the Lam Bar, Nurs-Ette, and Orphan Annie. Many growers make their own feeders that can range from a nipple mounted on the side of a plastic pail to elaborate plumbing systems with centrally refrigerated milk storage and supply.

Life as a bottle baby is pretty nice in a clean pen with a free-choice milk replacer and ground feed.

We use five-gallon plastic buckets that we get from a local bakery to make inexpensive self-feeders. We buy Lam Bar nipples and plastic tubes and mount them in the plastic buckets by drilling five-eighths-inch holes in the side about seven inches up from the bottom of the bucket and inserting the nipples. If the lambs chew on the nipples with their molars, they will quickly ruin them. This can be prevented by spacing the nipples about four inches apart and putting little metal baffles between them. Fold a tab on one edge of a three-inch square piece of sheet metal and attach it between nipples to the plastic bucket using screws. We have found that in some years lambs will chew nipples, in other years not.

If your lambing weather is so warm that spoilage of the milk or replacer is a problem, the nipples can be mounted on the side of a plastic foam icebox by drilling the holes in a strip of metal that is screwed onto the side of the icebox. Some people even get so fancy as to mount a strip of sheepskin, wool side out, behind the nipples to give a sheeplike touch to the whole apparatus. Spoilage can also be inhibited by mixing 1 cc of formalin (get it at your drugstore) to each gallon of replacer.

Lambs will have to be taught to use the self-feeder. Some learn immediately, though others take repeated guiding to the nipples by a patient, though sometimes frustrated, shepherd. If a number of lambs are in an orphan pen together, the slow ones sometimes learn from the others who have figured out how to use the nurser.

LABELING LAMBS

If you have just a few lambs, you may have no trouble telling them apart. If the numbers get larger, or if you are raising look-alike purebreds, you will have to mark them in some way. I'm a believer in the value of records, and that pretty much means giving each lamb a number. Naturally, lots of sheep and lambs end up with names too, but names don't enter into records as easily as numbers, especially if you are using a computer to keep track of your sheep.

One sort of temporary numbering that can be done is to use paint to mark a number on the side of the lambs and their mother so that they can be kept together. Usually the ewe and all of her lambs are given the same number. There are paints on the market, made especially for this purpose, that will mostly scour off when the wool is washed. Branding "irons" made of cast aluminum are available. The branders are dipped into the paint and applied to the wool. The paint-branded numbers are really handy when you find a crying lamb all by itself being summarily ignored by mother. With a number on it and on her, you can get them back together promptly. The disadvantage of paint branding is that a residue of the paint does persist on the wool for as much as a year with some sheep. Even though this special paint does scour out, no handspinner will welcome it, and even commercial buyers commonly pay less for branded wool. Believe it or not, another disadvantage of the paint brands is that they disappear from some ewes in a week or two. It seems to depend on the type of wool.

Whether or not paint branding is done, there should be some more permanent method of identification. A tattoo can be put inside the ear by using a special tool and tatoo ink. This is not often done with lambs because it takes a considerable amount of time, especially if many lambs are tattooed. With breeding stock it is a permanent method that is very useful for sheep with pale ears. The ink is applied to the inside of the ear first, then the tattoo device is used to puncture the numbers into the ear, and finally the ink is rubbed into the holes for a permanent ID mark. Do not overprint a failed try, as it will be unreadable.

Ear tags are a much commoner way of giving a sheep long-lasting identification. There are many types and brands of ear tags, and I won't attempt to be comprehensive but will point out a few principles.

For the lambs that are going to market, there is no point in investing a lot of money in fancy tags. The cheapest tag on the market as far as I know is the small brass type made by Ketchum. It is easily inserted using an inexpensive, special pair of pliers. The disadvantage is that it turns in the ear readily and can get to be very difficult to read without a lot of cleaning and twisting; this hurts the lamb so that it is struggling and fighting the whole time. Small Rototags are also good for lambs.

The next larger-sized tags, about an inch-and-one-quarter to an inch-and-one-half long, are made of aluminum or steel. These cost a bit more than the little brass ones, but they don't turn in the ear very readily. Large metal tags such as these are used by the breed associations to permanently mark registered animals. These tags are also inserted with special tools or with a common pair of pliers after a hole has been punched.

There are also metal tags with a circular portion that gives more space for information than a simple strip. This type is often used by research stations where the bookkeeping for a lot of different experimental flocks requires a lot of information on the tags. If you want to put quite a bit of information on a tag, don't forget that you can use the back side of the strip as well as the front. One can buy metal dies for impressing numbers and letters into tags, and the metal ones can also be painted for color coding.

Many different types of tags are made of plastic. These have the advantage of being available in a rainbow assortment of colors so that the color itself becomes part of the information. Plastic tags are handy, but they do have the disadvantage that the numbers wear off faster than on a metal tag. They also become brittle with age and exposure to sunlight and break. The sunlight damage can be minimized by placing them in the lower part of the ear to keep them shaded by the ear itself. Most manufacturers of plastic tags will imprint your name or the name of your farm on the back of the tag. Plastic tags require a punch or a special tool to insert them.

Finally, there are flexible rubber or soft plastic tags. These have an arrow-head-shaped piece that folds to pass through a small slit in the ear but then will not pass back through the same slit (like a barbed arrow). The slit can be cut with a knife or punch or there is a special tool (of course!) that makes the slit, folds the tag, and inserts it in one squeeze of the handles. These tags come already numbered or can be ordered blank and marked with special marking pens, whose marks the manufacturers claim to be permanent. They are big enough for shepherds with poor eyesight to read from a distance, or, alternately, lots of information can be crowded onto them—birth date, dam, sire, breeding, and the like—so the sheep is almost a walking record book. These tags, being made of soft, pliable material, have less of a tendency to tear out of ears than other types.

Lambs can be identified by notching their ears, using a notching tool such as is used for pigs. There are various conventions used for notching, but one that uses a relatively few notches for a big range of numbers employs a binary counting system. If you are not familiar with a binary counting system, it uses the sum of a series of numbers—1, 2, 4, 8, 16, 32, 64, 128, 256, 512, etc.—to indicate a given value. To give you an example, the binary number 11010 is the same as 26 because, reading the binary number from right to left, it means the sum of zero ones, one two, zero fours, one eight, and one sixteen equals twenty-six: or $(0 \times 1) + (1 \times 2) + (0 \times 4) + (1 \times 8) + (1 \times 16) = 26$. With only ten notch locations one can number up to 1,024 lambs. A notch can be used to indicate a one and no notch to indicate a zero. One notch at the tip of the ears and two notches each on the top and the bottom gives a total of ten locations. You will find that the lambs flinch quite a bit at each notching, so cutting up to ten notches may not be your cup of tea.

Some raisers use an ear notch or simply cut off the end of an ear to indicate that a ewe lamb is a twin or a triplet and should be saved as a replacement rather than shipped to market. Such a marking system is better than nothing, perhaps, but it does make the shepherd focus on only the fact that the ewe is one of a multiple birth. This may well be a plus, but the real question is not whether she was a twin or a triplet but how fast she grew and what sort of a track record her mother has, as an estimate of how good a producer she might prove to be. One wonders how many inferior ewes are retained just because the tip of their ear was cut off when they were born, when their productive, single-birth mates get culled.

The thoroughly modern shepherd can even implant ID chips in the ears for about seven to eight dollars a chip. You will need an electronic reader to use them. Such chips are used in some scrapie reduction programs. As with tattoos, such chips are for breeding stock, not lambs.

Whatever method you use, I urge you to label your lambs in some way: this will encourage you to keep good records of the lambs so you can identify your best ewes. Records make the long-term difference between profit and loss.

Detailed records are more difficult to keep in some situations. I refer to range flocks that are seldom located where tagging, weighing, and the like are very convenient. In days past, ewes had their lambs in the shelter of a sagebrush and took care of themselves and their offspring. In such a situation, a ewe that had a vigorous single lamb was actually preferred, and twin and triplet producers were culled to be sold to farm flocks or to market. Today, most range flocks are lambed out in shelters or yards of some sort and are eartagged and handled much as one would a farm flock. The big difference is that the flock moves around quite a bit on four feet, and a robust, single lamb will keep up better than a weaker pair of twins. A lost lamb is often abandoned or given away to a passerby because there is no way for the herder to locate its mother, and it certainly can't be raised on milk replacer on top of a mountain in Wyoming. Low lambing rates are tolerable with range flocks, since the investment per ewe is smaller than with farm flocks because of the low investment in facilities.

TAIL DOCKING

At the same time that ear tags are applied, a day or two after birth, the lamb's tails should be removed (docked). There are a few shepherds who have objections to docking lamb's tails. These people tend to be the ones who think that everything in life should be natural, and they cannot bear the thought of hurting the hapless little lamb by absconding with its tail. I will not question their motives or the genuineness of their sentiments, but I do think that their ideas bear examining. In the first place, a sheep is not "natural." Sheep have been domesticated for at least ten thousand years, and probably for a lot longer, and bear almost no resemblance to their wild precursors. They are a product of man and of civilization and should be treated as such. They need assistance from a shepherd, and that includes docking. An undocked lamb will accumulate feces under the tail if its droppings are anything but dry pellets, as is commonly the case during a lamb's first weeks. Except in midwinter, flies will lay eggs in the dung pad under the tail, and maggots will hatch out. When the maggots have consumed the feces, they will go after the lamb. You can't tell me that the maddening irritation and pain from maggots under the tail is better than a moment's distress when a tail is removed. Not only that, a plug of dung under a tail can stop up the lamb like a cork, and it will be very uncomfortable or worse.

There are breeds of sheep still raised in the Middle East and Asia that have very large, fat tails. These sheep are raised in part for the tail itself, which is considered a gourmet delicacy and a source of cooking fat. If you are a member of the ancient Samaritans who live still in Israel or of some other group that prizes the tail of a sheep that literally comes dragging its tail behind it, then, and only then, can I understand your not wanting to dock. Otherwise, I think docking is the humane and sensible thing to do. If an economic incentive is more meaningful to you, you should be aware that lambs sent to market are discounted a dollar a hundredweight (cwt) for undocked tails. In other words, you are docked for not docking, and you will probably get a poorer price for your lambs overall because the buyer considers tailed lambs a sign of poor management.

Docking a lamb tail is a quick and simple operation. I prefer a docker that is heated either by electricity (Meador's TNSC) or propane (Primus BJ5000). The heated cutter is used to sever the tail between vertebrae with a slow-cutting action. The tail comes away cleanly and the heat serves to cauterize the wound, which both reduces the risk of infection and reduces bleeding. To locate a place between vertebrae, just feel the tail with the fingers. A single vertebral bone is spool-shaped with ridges at each end. The space between two vertebrae is a ridge around the tail. Push the loose skin a bit toward the body of the lamb before you cut so there is some excess to cover the wound and speed healing.

Have someone hold the lamb for you, or hold it between your ankles, and dock away. If there is excessive bleeding, touch the hot cutter lightly to the bleeding place to seal it off. I like to avoid cutting too close to the rear of the lamb, and I tend to cut where the wool starts under the tail. Some advise removing the tail just to the tip side of the place where the two tendons that hold the tail come together in a V. No matter where you cut, do it between tail vertebrae. It is simple to feel where the spaces are. Show sheep used to be docked right up to the body for the sake of appearance. This practice is now widely frowned upon and not even allowed in some states. There is considerable evidence that short docking increases the risk of rectal prolapse.

If you don't like the idea of a heated docker, there are tools called Burdizzos and dockers that are used to crush and/or cut the tail. The Burdizzo crushes the tail, thereby restricting the flow of blood in the tail so that it atrophies and eventually falls off. The crushing-cutting type cuts off the tail and at the same time crushes the stub to reduce bleeding. Be sure to hold the tool correctly so you don't crush the cut-off portion instead of the stub.

Yet another way to dock a tail is to apply a small, strong rubber band with a tool called an Elastrator. The squeezing of the band cuts off blood circulation, and the tail eventually falls off. If one can judge from the behavior of the lambs, this rubber-band method seems to produce the most discomfort of the various methods.

The absolutely worst way to dock a tail is to cut it off with a sharp knife. Not that this is any more painful to the lamb, for it may well be less painful. The big disadvantage is that profuse bleeding often follows such a clean cut. Blood will squirt out in far-flying pulses, and bleeding from a severed tail is very difficult to stop. I know of shepherds who have lost lambs from bleeding to death as a result of tail docking. Letting a lamb die from improper docking is a spectacular example of poor management. Blood stopper may help reduce bleeding from any dock.

Infected Docks

Regardless of the method used, there is some risk of infection from docking. There is a bacterium of the genus *Corynebacterium* that is present around most barns and yards and readily enters a fresh wound such as a tail dock, shearing cut, or scratch. The electric docker is helpful in reducing infection but will not eliminate it altogether. The stub can be dipped in iodine as a preventive measure. Alternatively, the lamb can be given an injection of penicillin to prevent growth of *Corynebacterium*, which is especially sensitive to that antibiotic. If management and weather permits, the lambs can be turned out to a clean pasture where the risk of infection is practically nil. Do check the stubs to see that they heal. To treat an infected tail, thoroughly wash and clean it with soap and water, apply iodine, and give the lamb a penicillin shot.

Be sure to give the injection, *and all injections to young lambs,* subcutaneously (under the skin), even if the instructions on the bottle say otherwise.

Absorption of the medication will be a little slower, but you will avoid the risk of causing crippling nerve damage from an IM injection in a tiny lamb.

If *Corynebacterium* invades the body of a sheep it forms abscesses in lymph nodes and elsewhere, giving a disease called caseous lymphadenitis (CL) that causes wasting away of the sheep as more abscesses form in her body. There is no cure, but a new vaccine called Casebac or Casebac DT appears to be fairly effective. The strain of *Corynebacterium* causing CL appears to be introduced when an infected sheep is brought into the flock. If you have the problem, talk with your vet to set up a culling and vaccination program.

Tetanus

The other potential infection risk is from *Clostridium tetani,* causing tetanus. This microorganism, like all of the clostridia, is an anaerobe—it lives in the absence of air. There is little risk of *Cl. tetani* infection with the heated docker or cutting methods. However, such infections are fairly common with cut/crushing and even more so with Elastrator bands. Naturally, no bacterial infection will occur if the bacteria are not present. If you are not sure whether or not *Cl. tetani* is present on your property, you might elect to ignore the possibility of tetanus and treat cases as they arise, if they do arise.

If symptoms are detected early, treatment will usually be successful. Affected animals become paralyzed in the hind quarters. The tail may be stiff, ears erect, eyes staring, and swallowing and breathing may be affected. The animal may be startled by loud noises or sudden movements.

Treatment consists of large doses of antitoxin together with penicillin to reduce further growth of the bacteria. The site of infection should be cleaned out completely and washed with hydrogen peroxide solution. Some veterinarians will also administer tranquilizers during recovery. The treatment of an affected animal is usually not economically justified, however, since the medication costs more than the value of the lamb in most instances. We have saved some young lambs by treatment as above.

If *Cl. tetani* is known to be present on your property, prevention is much easier and cheaper than trying to cure sick animals. If the ewe is vaccinated prior to lambing, the lambs will have received protective antibodies from colostrum and will be able to fight off infection. With this method, the ewes should be vaccinated each year a few weeks before lambing. Vaccination of ewes is probably the cheapest protection if tetanus is a major problem for you. There are inexpensive vaccines that build ewe immunity for both *Cl.*

perfrigens (C & D) and *Cl. tetani.* If ewes were not vaccinated, one can give each lamb 100 IU of tetanus antitoxin at docking or castration, but that is more costly.

Flies

Finally, the dock should be protected against flies if it is fly season. Flies will lay eggs in any dead tissue, and the maggots will cause irritation and even death in severe infestations. Try to avoid fly season by early docking. If your lambing schedule makes that impossible, spray the stub and surrounding areas with a fly repellent. Some sprays contain both a repellent and an insecticide to take care of the individual flies that ignore the repellent.

CASTRATION

Just as there is disagreement about docking of tails, there are all shades of opinion about castration of male lambs. Many male lambs that are destined for slaughter have traditionally been castrated, that is, converted to wethers. The commercial lamb markets are set up on this assumption, and ewe and wether lambs are generally bought at the same price. Depending on the season of the year, the part of the country, the buyer, and a host of other factors, ram lambs can command the same price as ewes and wethers, although more typically they are discounted for their maleness. Discounts vary, but can be substantial. In the American market, ram lambs are lumped with ewes and wethers until early to mid-summer, at which time a discount of two or more dollars per hundredweight is applied to finished ram lambs being sold for slaughter. Lambs sold as feeders can be discounted much more severely for being rams because of the nuisance they cause the feedlot operator.

The packers object to uncastrated lambs for a number of reasons. As uncastrated lambs get older, they gradually develop male characteristics that the packer doesn't like. As maleness becomes more evident, weight is added at the shoulders rather than the more valuable loin and hind legs. The skin becomes more difficult to remove over the shoulders, which costs time at the slaughterhouse. Furthermore, the thin membrane covering the meat may be torn, which results in extra weight loss from the drying of that part of the carcass. With older uncastrated lambs, the fat becomes yellow in color instead of white, and there is resistance to this at the retail level. There is also, of course, the weight of the scrotum and testicles, which have little market value in some areas.

Lamb feeders do not want uncastrated lambs because their sexual activity in chasing and trying to mount ewe lambs and one another disrupts the peace and quiet of the feedlot, with resulting reductions in rates of weight gain. Some feeders have facilities to house males separately and will tolerate them, but they still will not pay as much for entire rams as they will for wethers.

There are also disadvantages to the shepherd from not castrating. Like the feedlot, the pasture or barnyard will be a lot more active with uncut males chasing around. Worse yet, some of the uncut rams may breed some of the early-maturing ewe lambs. Even if the ewe was one that you wanted to keep as a replacement for the flock, you would want to breed her to the ram of your choice at a higher weight and later age. Breeding a youngster to a scrub ram lamb is no way to improve the flock.

With all of these considerations indicating that castration is in order, is there any reason even to consider not doing it? Yes, there is. The big reason to avoid castration is that it causes a setback in the growth of the lamb at the time of castration. In a business where fast growth on the least feed spells the difference between profit and loss, the slowing of growth at castration can be significant. In addition to the reduction of growth for a short period, entire rams grow faster on average than their castrated counterparts, all else being equal. If a ram lamb grows 10 percent faster on the same feed consumption, but is then docked 5 percent at marketing, you are still ahead of the game relative to making wethers. Also, if you count labor as a cost, the time and expense of castrating is eliminated altogether by keeping ram lambs entire.

In a study by A. L. Slyter at South Dakota State University, growth of rams and wethers was compared. One group was weaned at 90 days and put on feed, a second group was weaned at 120 days and finished on feed, and a third group was left on pasture with the ewes. Over two years of trials, the ram lambs averaged 11.5 percent, 4.6 percent, and 3.2 percent heavier than wethers in the three groups. The grand average for all test groups showed ram lambs gaining 6.4 percent over their wethered cohorts. If lambs were worth $60 per cwt, the discount for rams would have to be almost $4 to favor wethers. For the group weaned at 90 days, it would have to be nearly double that. Also, interestingly enough, over the two years of the trials, the wethers ate about 20 percent more feed in the feedlot, although gaining less weight than the rams.

With some flocks, castration is avoided because the shepherd hopes to sell rams as breeders. This is a judgment call, because only the breeder's past experience can tell him how many rams he can expect to sell. If wishful thinking overwhelms realism, the ram breeder may end up with a lot of rams that he doesn't have a buyer for and will take a beating at the slaughter market.

One compromise is to castrate some and not others. Any rams that can be expected to reach market weight and finish before discounts are applied in July or so can be left entire. Later lambs can be castrated to avoid the dock for entire rams.

Yet another possibility is to turn the rams into cryptorchids or "crypts," also called short-scrotum rams. A cryptorchid is a ram with the testicles remaining in the body cavity instead of down in the scrotum. Some rams are born this way, and the testicles never descend or are pulled up later. Most crypts are man-made, however. The usual method is to use Elastrator rubber bands. The testicles are pushed up into the body, and the band is applied to the scrotum. The scrotum falls off eventually, leaving the testicles in the body cavity. When the testicles are pushed up, the shepherd should hold the surrounding tissue so the attachment of the testicle to the inner part of the scrotum (the tunica vaginalis) is pulled loose. Also, the band should be placed as close to the abdomen as possible to do a neat job. Crypted rams are usually sterile because the sperm cannot tolerate the high temperatures inside the body. Insofar as hormone production is concerned, the crypt lamb remains the same as an entire ram lamb and will show male body conformation and respond sexually as a male. Thus, cryptorchid males will create the same uproar in your barnyard as a ram lamb, but at least they won't get the ewes pregnant. If you do crypt your lambs, in all fairness to the feedlot operators you should finish them yourself rather than foist them off as wethers.

The big plus for cryptorchid lambs is that they grow at least as well as entire ram lambs, and there are reports that they grow better. Also, they will be treated as wethers at most markets, even when penalties on rams are applied.

The decision whether to castrate is one that only you can make. You should check with local buyers to see what the market demands. Some markets will accept only ewes and wethers and won't take ram lambs at any price. Other places don't care. Some of the packing houses don't want rams at all and wouldn't buy any if they thought they could get away with it. There is still enough competition, however, so that they cannot dictate completely to the shepherd.

An important exception is the increasingly important Muslim market. Most Muslims prefer entire rams, and will pay extra to get them. If you can develop a local market among Muslim customers, you will, of course, not castrate. There are even some packing houses now buying for the "ethnic" market that have no objection to entire rams.

Many raisers in our area leave January and February lambs as rams and castrate later ones in order to accommodate market demands. We mostly castrate except for purebreds and exceptional crossbreds that we reserve for sale as breeding stock. When we have crypted, we have never been docked for the crypts, but our crossbreds are of a breed that is slow to show heavy masculine shoulders and neck. Ram lambs that show their maleness early probably should be castrated unless you are willing to take a chance on getting a higher total net profit even with a ram discount at the marketplace.

Castrating is done in a variety of ways, as you might imagine. The testicles can be removed entirely by pushing them up into the body, cutting the scrotum at its widest part with a sharp knife, then pulling the testicles until the cords holding them break. Pulling them is easier said than done, because they are slippery, and getting a grip is not always simple. Some shepherds use a toothed, plierslike tool to grasp and pull the testicles. The traditional way is to grasp them with the teeth, pull, and spit them into a waiting bucket. Usually one person holds the lamb and the other cuts and pulls. The severed testicles should be saved for a gourmet treat of "mountain oysters." Remove the outer covering and fry them or prepare them as you would sweetbreads. If the idea doesn't put you off, you'll find that they are a treat.

Elastrator bands can be placed over the whole scrotum, including the testicles, to pinch off the cords to the testicles and the scrotum. In some lambs, the testicles may not have descended into the scrotum, so push with your fingers on the abdomen in front of the scrotum to pop them into the sack, and then apply the band. Take care not to pinch one of the little teats in the band, and make sure that both testicles are fully into the scrotum and not caught partly by an Elastrator band.

A Burdizzo can also be used to crush the cord leading to the testicles, resulting in a loss of blood supply and atrophy of the testicles. Crush each cord twice to double your chances of getting the job done successfully. Have a veterinarian or someone who is experienced demonstrate the use of a Burdizzo to you because it won't work if it isn't done correctly. This is the preferred method for older rams.

As to when to castrate, the answer is as soon as possible. Very young lambs seem hardly to notice that anything has happened to them, although they may walk stiffly for a few days. The longer you wait, the harder it is on them and the bigger the growth setback they will suffer. Try to do them while they are still in the jugs, no later than day three, and you will cause fewer problems for both the lamb and yourself. If the lamb is weak or sick, castration should be put off until it is healthy and strong. The possibilities of infection are the same as for docking, so do both procedures at the same time and take the necessary precautions. If you cut the scrotum, be sure that the wound drains and does not seal up and become infected. Open and clean any scrotums that seal over.

Medical

The Ewe

The principal cause of concern for the ewe at lambing is damage to or infection of her uterus and vagina. With a normal, unassisted birth there is little likelihood of any problem, but if the shepherd or veterinarian has to insert a hand and arm inside the ewe to assist in lambing, there is always the danger of damage or infection. Damage can be minimized by being sure that your fingernails are short and by being very gentle and careful when working inside the uterus. You may have to use considerable force to pull a lamb that is in an awkward position, but that doesn't mean using abrupt movements or violent yanks and twists. Just keep calm, work slowly and carefully, and damage is unlikely.

For cleanliness, wash your hand and forearm or wear gloves of some sort. Veterinarians commonly use a shoulder-length rubber glove to shield their whole arm, but the average shepherd doesn't generally have those. A good scrubbing, including under the fingernails, will usually suffice.

There are boluses—usually made of a combination of the chemical urea and one or more of the sulfonamides (sulfas)—that are intended to be placed in the uterus after an assisted birth or after an inspection with the hand and arm. Our vet thinks that these are of value only to help the peace of mind of the shepherd and do nothing at all for the ewe.

After an assisted birth we commonly give an injection of an antibiotic to help ward off infection, especially if a lot of work was needed. Some authorities recommend an injection of oxytocin to contract the uterus and ward off uterine prolapse, as well as to help in expelling afterbirth tissues. Follow your veterinarian's recommendations.

If a ewe has a vaginal prolapse after lambing, she can be treated as described in Late Gestation. Be sure to treat the cause of the prolapse, such as infection, because unlike the vaginal prolapses prior to lambing you can't blame this on lack of room inside the ewe. Some authorities say to cull any ewe who prolapses after lambing. We have kept ewes who prolapsed after their first lambing, and they have never done it again.

The big problem that arises occasionally is that not only will the vagina prolapse, but the entire uterus will evert. It is a large red body of tissue that has knob- or buttonlike structures (cotyledons) over its surface, and it will hang from the prolapsed vagina in most cases. If you want to try to push it back in yourself, wash the whole mass with a mild detergent and warm water. If it is really swollen, sprinkle a pound of granulated sugar over it to help shrink it for easier replacement. Hang the ewe by her heels, or have an assistant hold her, lubricate the uterus and push it back in carefully. When it is replaced, be sure that it is completely back, and not still partly inside out as it was when it was outside of the ewe. If it is not completely back in a normal configuration and position, it will come right back out again. In fact, it may come out even if you get it in correctly.

Once the uterus is back inside the ewe, give her an injection of oxytocin, suture across the vulva to keep everything inside, and put on a twine harness as described in Late Gestation. Then pray or cross your fingers. Ewes frequently die from internal bleeding after a uterine prolapse, but you should try the rescue operation anyway. You really have nothing to lose, after all. If you prefer not to try this sort of major operation, check with your vet on the fee because the rescue may cost more than the value of the ewe.

Be alert for any signs of milk fever as described in Late Gestation. This is not common with sheep, but heavier milking breeds and individuals are susceptible, so be alert. It is easy to treat when caught early.

The Lambs

The newborn lamb has left the comparative safety and protection of its mother's uterus and is exposed to the dangers of the world. In a lambing barn the shepherd has removed physical dangers such as predators and cold temperatures, but there are still biological dangers.

The lamb is subject to infection by a myriad of pathogenic organisms. Its defenses against invasion by these freeloaders are complex, but the immune system figures prominently. A newborn lamb is capable of having an immune reaction against an invading organism as soon as the lamb hits the ground. It is said to be immunologically competent, in the sense that all parts of the immune system are present and in working order. However, the immune system needs time to react because even though the immune system is present, building immunity to any given organism takes at least a couple of weeks, so for the moment the little one is essentially unprotected.

The immune defense that the lamb can use right away is the chemicals called antibodies. It receives these from the ewe in the colostrum it drinks during the first couple of days of life. Antibodies are large protein molecules that are able to attach themselves to specific bacteria, viruses, or other microorganisms in the blood or in the lymphatic system and either inactivate them or aid in their destruction. There are also antibodies that attach to and neutralize toxins that may be produced by growing organisms. The temporary immunity given by the antibodies from the ewe's colostrum is called passive immunity. It lasts only as long as the unused antibody is present in the lamb's body, because the lamb does not replenish the used antibody automatically.

Antibodies are very specific in their action; they are able to attach themselves only to a single type of microorganism or toxin, or in some cases to very closely related materials. The antibody has a cavity at one end that is made of amino acid building blocks in a unique configuration that will link only with a matching projection on an object such as a bacterium. Thus, if the antibody is effective against bacteria A, it will be ineffective (or at least not totally effective) against bacteria B. Antibodies are formed in response to proteins called antigens that are unique to each type of organism. It is sometimes said that each organism or other object has its own characteristic antigenic signature.

An antibody that is specific against bacteria A was originally formed in response to the presence of some of bacteria A in the sheep's lymphatic system. Thus, the ewe's colostrum will contain antibodies against every antigen that her immune system has ever encountered in her entire lifetime, which means against every pathogenic organism that has ever infected her, whether it made her sick or not. For this reason, colostrum from an old ewe is better protection for a lamb than that from a young ewe, because the old one will have encountered more antigens in her longer lifetime than the youngster.

The Immune System

Armed with a mother's antibodies, the lamb is able to control attacks by pathogenic organisms. At the same time it begins to build its own defenses as its immune system encounters antigenic material. The immune system is highly complex, and research in this field has been rapidly expanding over the past decades, but some generalities are agreed upon by immunologists. First of all, there are key parts of the immune system that are able to distinguish between self and nonself tissues. The immune reaction is against nonself material of any kind, whether it is a virus or an intestinal worm or a toxic molecule.

There are two tasks accomplished in an immune reaction. One of these is to form antibodies that attach to foreign bodies, and the other is to attack them with special sorts of white blood cells. These two types of response are called antibody-mediated and cell-mediated. Both responses have their origin in the bone marrow, where the so-called stem cells are manufactured. These develop into lymphocytes and macorophages. The lymphocytes subdivide into one population that passes through the thymus gland to emerge as T cells, and one population that does not, called B cells. The macrophages also split into two groups, some that attack and eat foreign bodies and some that process antigens for the other cells.

The four types of cells act together to give a broad attack on invading antigenic bodies. Macrophages attach antigens from the nonself material to themselves and pair them with other proteins called immune genes or I genes. The pair, consisting of an antigen and an I gene, is presented to T cells of a type called helper-T cells (T_H cells). The T_H cells then carry the antigenic message to other T cells called killer-T cells or effector-T cells (T_E cells) as well as to B cells.

B cells that get the message from a T_H cell change into large cells called plasma cells that generate large quantities of antibody that is specific for the antigen delivered by the T_H cell. Some of these antibodies, called immunoglobulin gamma or IgG, enter the blood and lymphatic system and attach to and neutralize any foreign bodies that carry the identifying antigen. Others, such as the delta type, IgD, remain on the surface of the cell that produced them and do their work there. Yet others, like IgE, function to fight internal parasites such as worms but can also trigger harmful allergic reactions. There are many more types of immunoglobulins.

T cells that get the T_H message perform two main functions. The killer T_E cells attack and destroy cells that bear the identifying antigenic signature.

T cells produce substances called lymphokines that have wide effects. They convert some macrophages into phagocytic (cell-eating) cells. Lymphokines also attract leucocytes and macrophages to parts of the body that are under attack by nonself organisms or other material and cause an inflammation as the foreign bodies are destroyed or inactivated.

Once T and B cells have been given the antigenic signature of a particular antigenic substance, they retain this information. The T and B cells are also very long-lived cells, so the "memory" of a particular antigenic object is carried in the sheep's body for many years. When one of the coded cells encounters its own special antigen, it responds by enlarging into a large cell called a lymphoblast that then subdivides into numerous exact copies (or clones) of the original cell.

In the case of T_E cells, the immune response is very rapid. Each T_E cell that encounters its special antigen is very rapidly formed into a huge number of additional T_E cells that go about this business of attacking the antigenic material, forming lymphokines, and so forth.

Response of the B cells is somewhat slower because they first have to go through the process of cloning; then the clones have to form plasma cells that actually make the antibody that attaches to the antigenic material. It is the relative slowness of this process, a matter of a couple of weeks or so to build high antibody levels, that makes the passive immunity from the ewe so valuable to the lamb. The presence of the passive antibody gives a grace period during which the lamb's immune system can make its own antibody and acquire active immunity.

There is one hitch in this whole system that may have occurred to you by now. A lamb cannot build active immunity to an antigenic material until it has been encountered by a macrophage (actually many of them) to start the whole process of the immune response. As long as the lamb's active immunity is built while it still has passive immunity from the ewe, all is well. If the lamb is attacked by a pathogenic organism when it has neither passive nor active immunity, it may well get sick or even die before the immune system can respond. If it survives the attack, it will be able to respond rapidly to future invasions, however.

Immunization
This is where vaccination or immunization enters the picture as a practice to prevent disease. Vaccination is named from the name for the cowpox virus,

vaccinia, which was used to give immunity to smallpox. People who had been infected with *vaccinia* were found to be immune to the smallpox virus, and intentional infection with *vaccinia*—that is, vaccination—was used for many years before anyone understood why it worked. *Vaccinia* is either a mildly virulent form of the smallpox virus or has a similar enough antigenic signature to the smallpox virus that the immune system responds to the smallpox virus after "learning" the cowpox signature. The term "vaccination" gradually came to be used for any immunization against a virus and is now generally used loosely to mean any immunization against a virus, bacterium, chlamydia, or other pathogenic organism. What all vaccines have in common is the ability to prepare the immune system for rapid response to an invading pathogenic organism without actually causing the disease.

For example, a bacterium or virus can be killed or chemically weakened so that it cannot multiply and cause infection or disease. The killed or weakened organism will still carry the antigenic message on its surface, however, so it is still capable of stimulating the immune reaction, even though it causes no illness. Organisms can also be weakened in other ways so as to make them harmless. This is done for some viruses by growing them on the tissue of a different species than the one to be protected. For example, a virus can be grown on horse cells in such a way that it becomes very virulent for horses, but loses its ability to attack sheep cells. This attenuated virus can then be given live to a sheep to build immunity without making the sheep sick. Or, the poisons produced by a pathogenic organism can be chemically altered to be nonpoisonous, but still carry the same antigenic signature so that immunity to the toxin can be built without actually poisoning the animal.

Lambs can acquire immunity to common sheep diseases through appropriate vaccinations. Then their immune system is prepared to react quickly if the specific organisms for which it is immune are recognized by the patrolling *T* and *B* cells.

Sheep are vaccinated for some common lamb diseases prior to lambing so that antibody levels in their colostrum have time to build to high levels at lambing. Thus, the ewe is vaccinated for the protection of the lamb yet to be born. Also, of course, both lambs and adult sheep are commonly vaccinated for their own protection.

Vaccinations are effective as long as there remain a significant number of the appropriate *T* and *B* cells still present in the sheep's body. In order to keep their numbers high, annual or less frequent boosters are given for some diseases. For other diseases, a single vaccination appears to confer lifelong immunity. Recommendations for different diseases and different types of vaccine are not all the same, so be sure to read labels and consult with your veterinarian about your immunization program. It is important to give the recommended dose of a vaccine. Do not reduce the amount for lambs. Generally, all vaccines are given in two doses, with the second one two to three weeks after the first. Insufficient vaccination can invite disease. On the other hand, too frequent vaccination costs a lot of money, can trigger harmful side effects, and should be avoided if it is not really needed.

Vaccines, Etc.

There are a number of different preparations that are used for immunological treatment or prevention of disease. In current usage a vaccine means any material that can be used to stimulate an immune response without giving a full-blown case of the disease. Some vaccines, such as sore-mouth vaccine, are actual live viruses of full virulence. Others, such as various types of rabies vaccines, are viruses rendered noninfective by some means. Still others are killed bacteria called bacterins, such as those used to vaccinate against vibrio. Some are inactivated toxins called toxoids, such as those used to prevent enterotoxemia caused by *Clostridium perfringens* toxin. There are also materials called antitoxins (these are not the same as toxoids) that are mixtures of specific antibodies that attach to and inactivate a given toxin or set of toxins. An injection of antitoxin gives short-term passive immunity in much the same way that sucking colostrum gives passive immunity to a lamb. If an active case of the disease is present, antitoxins can be used as a therapeutic measure along with other treatment to control growth of the pathogen. Antitoxins are commonly used against tetanus and enterotoxemia.

The question sometimes arises as to whether a toxoid and an antitoxin can be administered at the same time. After all, won't the antitoxin just neutralize the toxoid? Yes, but not until the toxoid has had some minor effect. However, the beneficial effect of the toxoid in building immunity will be greatly reduced.

Another question is whether antibiotics can be used at the same time as an immunological treatment. You bet they can. In fact, without participation of an animal's own immune system, fighting off sickness with just antibiotics is pretty much useless. Antibiotics are great, but what they really do is to sufficiently inhibit the growth of pathogenic organisms so that the immune system can do its job.

The body has other defenses against invasion by pathogens such as the skin, active membranes in the mouth, nose, and other openings, tears in the eye, sweat, and so forth. The immune system does not protect against everything. Pathogenic organisms can live in the intestine of a sheep and cause neither disease nor an immune reaction. For that matter, consider that a lamb is nonself to a ewe's immune system, yet the fetus is not destroyed by an immune reaction from the ewe because the fetus is protected. Also, some pathogenic organisms have coatings that inhibit recognition of their antigenic signature, so they largely escape the immune response. Yet others change their character rapidly to escape detection by the immune system.

Chemotherapy

Chemotherapy is a general term meaning medical treatment using chemicals (what most of us would call drug therapy). The sulfonamides or sulfa drugs were the first major group of drugs discovered that could be used to help the body fight disease—the first of the so-called miracle drugs. They function by inhibiting paraminobenzoic acid, an amino acid that many microorganisms need to grow and multiply. The sulfas don't actually kill the microorganisms that are producing a disease; they inhibit their growth until the lymphocytes act to destroy them. Sulfas are mostly excreted in the urine and can cause kidney damage if the animal is not provided with sufficient water. In using sulfas be sure to follow label directions for length of treatment so as to limit adverse side effects. Some sulfas are not readily absorbed and therefore act mainly in the gut if given orally (sulfasuxidene, sulfathalidine, sulfaguanidine) whereas others are readily absorbed (sulfathiazole, sulfamirazine, sulfamethazine). Don't give a sulfa unless you are sure you are using it for the right purpose. Check with your veterinarian to be sure.

Various compounds produced by microorganisms and their synthetic analogs are called antibiotics, the best known of which is penicillin. Some types (penicillin, bacitracin) are bactericidal—that is, they kill the bacteria or other organisms, principally by effects on the cell walls or cell membranes. Others are bacteriostatic in the same way as the sulfas, in that they affect protein synthesis and retard growth rather than actually killing the organisms (tetracyclines). It is not advisable to use a bactericidal antibiotic together with a bacteriostatic one because their actions do not complement one another.

Penicillin comes in a vast variety of forms, and your veterinarian may have good reason to prefer one type over another for any given health problem.

The common over-the-counter type is penicillin G, usually with procaine to slow absorption. There are also quick-acting types and others that are released very slowly to reduce the need for frequent treatment. It is very rare that a single treatment with penicillin or any antibiotic is called for. In most instances a series of doses is given, and the whole recommended series should always be given. Also, be sure always to give the full recommended dose, because a smaller dose has the effect of helping the microorganisms build immunity to the drug. Certain species of the genus *Streptococcus* developed the ability to secrete an enzyme that destroys penicillin, and this penicillinase-producing gene has appeared in other bacteria, by transfer from the strep species that "developed" the gene, much like an inventor licenses a process to manufacturers. Giving full doses and the whole course of treatment helps limit development of resistant strains with the penicillinase gene. Penicillin is mostly effective against a class of bacteria called gram positive and against chlamydia.

The tetracyclines include chlortetracycline (Aureomycin), oxytetracycline (Terramycin, Oxy-tet), and plain tetracycline. The commonest one is oxytetracycline as the hydrochloride. It is readily absorbed from the intestinal tract and is also given by injection. Even by injection it can have adverse effects on the rumen microflora. The action is bacteristatic. Many veterinarians prefer tetracyclines for respiratory diseases. Because of their ability to be readily absorbed in various parts of the body, they are also used to treat infections of the brain that are unreachable by most medications because of natural barriers.

There are long-acting versions of both penicillins and tetracyclines, which means one does not have to catch and treat the animal as frequently. A long-acting tetracycline called LA200 is widely used.

The shepherd should realize that no antibiotic has any effect whatsoever on viruses or fungi. In some viral diseases, such as sore mouth, antibiotics are sometimes given to help prevent secondary infection by bacteria. Antibiotics are only effective if the right one is used at the right time in the right dosage. This usually means using them on the advice of your veterinarian unless you have had a great deal of experience and can recognize a given condition for which you know the appropriate drug and dosage.

There are many antibiotics and other drugs that are available only from veterinarians or by prescription, and those may be the drug of choice, so don't just use what you can buy off the shelf at a farm supply store because you think you'll save a few dollars. You may lose the animal or fail to stop an epidemic in a flock by inappropriate treatment. Be cautious until you really know what you are doing.

Early Vaccination

After that little side trip into immunology and chemotherapy, let's get back to the lambs. Many producers like to vaccinate lambs against various diseases soon after birth. Tetanus protection can be given by injection of antitoxin, and protection against enterotoxemia can be assured by giving *Cl. perfringens* type C & D antitoxin. If the lamb received colostrum from a ewe who had been given boosters against these diseases, that additional protection should not be needed. Vaccination with *Cl. perfringens* toxoid or bacterin can be safely put off until the lambs are three months old—if the ewes were given boosters, and the lamb received adequate colostrum.

If your flock has immunity to soremouth, then the lambs will also have protection against that virus from their mother's colostrum.

If a lamb received no colostrum at birth, it should at least be given an injection of antitoxin against *Cl. perfringens* type C, and probably *Cl. tetani* as well.

We have vaccinated lambs at birth against the parainfluenza-3 or PI-3 virus, using nasal vaccine on a trial basis as recommended by a number of veterinarians. This virus can cause an infection of the windpipe (trachea) that leaves the lamb without some physical defenses against invasion of the lungs by other infectious agents that can cause pneumonia. There have been no controlled trials with sheep and lambs, and there is no proof that the PI-3 vaccine does any good at all, but clinical experience by vets who work with lambs a lot looks promising. The first year we tried it we had a greatly reduced incidence of pneumonia as compared to previous years. This proves nothing, because every year is different from every other, but we continued to use it until we felt that ewe immunity was adequate. The nasal vaccines are not approved for sheep, and therefore the package gives no instructions for use with lambs. Two well-known sheep vets recommend 0.5 cc up one nostril at birth, and that is what we used.

The PI-3 nasal vaccine (Naselgen, Aeromune) is combined with a vaccine against IBR (infectious bovine rhinotracheitis), which is not at all common in sheep, but a little extra immunity won't hurt. Ideally, all lambs should be given a follow-up treatment in four to six weeks.

Selenium and Vitamin E

If white-muscle disease has been a problem in your flock in the past, lambs can be given an injection of a commercial selenium–vitamin E mixture that you can obtain from your veterinarian. This mixture was developed for the selenium-deficient Pacific Northwest. It does not contain enough vitamin E for some areas of the country that may have sufficient selenium or a modest deficiency but that have vitamin E deficiencies in feedstuffs for some other reason. In informal clinical trials, G. E. Kennedy, a vet from Pipestone, Minnesota, has found that giving additional vitamin E is helpful to prevent white-muscle disease in that area. Dr. Kennedy sells some products by mail order that may be useful for white-muscle treatment. See Pipestone Veterinary Clinic in appendix 6.

Iron

According to many "experts," sheep and lambs have no need for supplemental iron in their diet. Some formal studies in Canada and some barnyard experiments suggest that iron supplementation is very helpful indeed. We had always had a problem with lambs eating dirt when they got access to the barnyard. The dirt eating caused constipation problems, and finally one year a lamb died from the constipation. I cut it open after it died and found that its true stomach (the abomasum) contained a large handful of dirt, firmly packed. It had died of starvation because no food could pass this dirt plug. I suspected that iron deficiency might be a contributing cause of the appetite for dirt. We immediately gave the entire lamb flock injections of iron dextran at a rate of 1 cc per lamb, and the dirt eating ceased abruptly. The fact that providing iron stopped the dirt eating does not prove that an iron deficiency existed, but to be pragmatic, if you observe dirt eating you might want to consider trying the injectable iron compounds used for piglets. Ewe milk contains virtually no iron at all, so a lamb that is not eating any solid food has no dietary iron source.

Entropion

Some lambs are born with an eyelid turned into the eyeball. This condition, called entropion, causes lots of irritation of the eyeball from the lashes rubbing against it. Left alone, a few cases will take care of themselves, but most will require some sort of treatment. Treatment can be as simple as rolling out the eyelid a few times, after which it may stay out. If that fails, try to dry off the lid and surrounding area enough to get some adhesive tape to stick and tape the lid out. If the eyeball is irritated, apply some opthalmic ointment to soothe it or use a mastitis ointment of the type intended for injection into cow's teats.

If these simple solutions don't work, you can take a surgical route. Many veterinarians repair the condition by removing an elliptical piece of tissue from the lid and sewing the edges back together to shorten the eyelid and keep it away from the eyeball. Check with your vet on the price for this surgery, because it may be a bit high in relation to the value of the lamb.

A surgical technique used by many producers is to nick the eyelid with a scalpel or razor blade, making one or two short cuts in the lid, parallel to the edge. These nicks should not go through the lid but just cut the outside. On healing, the tissues shrink at the cuts and hold the eyelid taut and out of the eye. Either along with the nicking, or as a separate treatment, a single stitch, using a needle and thread, can also be made to hold the eyelid away from the eyeball. Metal surgical clips can also be used.

The method we prefer was invented by veterinarian Robert G. Stewart of Burbank, Illinois. He felt that the surgical procedures were too complicated, so he developed a nonsurgical technique that anyone can use. First, the eyelid is injected with an antibiotic preparation to inflate it and to pull the inverted edge away from the eyeball. Second, the eyeball is treated with a topical application of antibiotic.

Stewart first used oil-based, long-acting penicillin to inflate the lid. This worked fine, but the product has been discontinued by the manufacturer. Stewart suggests substituting an oil-based teat infusion, available from veterinarians, called Quartermaster. Either transfer the material to a sterile bottle or draw it straight from the infusion syringe by means of a hypodermic needle. To inflate the lid, use a short, fine needle such as a half-inch, 20- or 22-gauge. Hold the lamb's head firmly and insert the needle just under the skin of the lid, about three-sixteenths of an inch from the edge of the lid. Inject about 1 cc of the antibiotic. If you are nervous about injecting an eyelid, get your vet to show you how. It is a good trick to know for pinkeye treatments too. Dr. Stewart does not recommend the use of presently available long-acting penicillins because he feels that they may be too irritating. However, I have used a long-acting penicillin without any apparent irritation.

In the second step, the eyeball should be treated with an antibiotic ointment of some sort that can even be the same medication that was injected into the eyelid. Note that this is a topical treatment and is not injected into the eyeball.

OBSERVATIONS

The shepherd should be alert for various problems both ewes and lambs may have around lambing time. Make frequent checks of the drop band, the ewes and lambs in jugs, and those that are already out of the barn. Time spent in checking will be rewarded with fewer troubles that get out of hand.

The only way a shepherd can expect to spot a ewe having trouble in lambing is to have watched a lot of ewes who did not have trouble. If normal behavior isn't known, then one who needs help won't be spotted. If you are brand new to sheep, try to spend some time with an experienced shepherd at lambing before you are on your own. Alternatively, get on good terms with a sheep raiser whom you can talk to over the phone about things you don't understand. Even without any help or experience, just use common sense, observe a lot, and you'll get along. You'll make mistakes, but you'll learn.

In a perfect world, all lambs are born active and ready to go. After a few minutes of rest they'll be up looking for the teat and ready to get on with life. Many lambs will do just that, especially crossbreds with a little Finn in them, though some are just not ready to get going. If you are there, and you should be, a weak or slow lamb can be given a meal of colostrum by a bottle nipple or with a stomach feeder. In fact, it is a good idea to give each of a set of triplets, or even twins, a bottle meal just to get them used to what a rubber nipple is in case they have to be supplemented or be transferred to a self-feeder for some reason. If a lamb knows only the real thing, it will view a rubber substitute with great disdain.

If you weren't there at the birth, you won't know whether or not the lamb got a meal from mom. Observe the lamb, and you'll probably catch it nursing. If it doesn't seem to be nursing, watch out. If it won't rise or if it stands humped up, perhaps shivering, it may not have had its first meal. Remember to stick a finger in its mouth. If the mouth feels cold to the touch, the lamb needs a meal of colostrum and should be fed in some way.

Sometimes lambs will act as if they had stiffness in the hindquarters and they may cry out. There are many possible causes for this, but don't overlook ordinary constipation. Lambs are born with some feces ready to be expelled, and in fact you will see some births in which the fluids and the lambs themselves are dotted with dark brown pieces of feces that were expelled when the lamb was still in the mother. This first dung, usually a dark brown to black, tarry material (called the meconium) must be gotten rid of within the first hours after birth. Retained meconium causes the bowel and intestine to fill with gas—a condition called rattle belly in Britain, where it is common in some areas, because the lamb makes a rattling sound when picked up. Another symptom is "watery mouth," which is just what it sounds like. If the meconium has not passed, an enema of soapy water or an oral dose of a teaspoonful or so of mineral oil will alleviate the condition. If you see mustard yellow to orange feces, the meconium has already passed, and all is well.

Diarrhea (scours) in very young lambs is rare, but not unknown. The sticky yellow feces is normal, but if it is loose and runny it is probably that nasty anerobe *Clostridium perfringens* growing in the gut, producing toxins and irritation. A single oral dose of soluble Terramycin given with the little rubber-bulbed dropper that comes in the package will generally clear up these very early scours. Others may prefer other drugs, but you'll find that tetracyclines of some type are the favorites.

Be on the lookout for injured lambs. Ewes sometimes get overenthusiastic in the cleanup procedures and chew off tails or ears, or they may paw at the lambs so much in an attempt to make them get on their feet that they will injure them. Some ewes, especially first-time mothers, may reject one or more lambs and injure them by hitting them with a head or nose. These ewes are best tied, or the lambs given a hiding place in a corner. Some ewes are just plain oafs and will sit on a lamb and injure it or even kill it. For these cases, you should use a larger lambing jug—a klutz jug as we call it—to give the lambs a chance to escape being flattened by their mother. Alternatively, they too can be given a hiding place in a corner.

Keep your eye on the ewe for the first day or so. She should pass the placenta within the first twelve hours or so after lambing. If she doesn't, you might want to consult your vet, because retained placentas can cause serious infections. Most ewes birth the placenta with no trouble at all.

Check the ewe's bag at birth and after. A normal ewe will have a full, yet pliable, bag that produces a good supply of milk. If the bag is really swollen, and full of milk, use this opportunity to steal a little colostrum for reserve and relieve the pressure at the same time.

If the ewe has insufficient milk, give her lots of water and good feed, and supplement her youngster's diet with some colostrum you stole from an overachiever. Try oxytocin to help her let down, although a dairy farmer would tell you to get down on your knees and massage her bag with a warm towel to help get things going. Once she starts to produce, don't supplement the lamb's food, because she needs frequent sucking to stimulate her own milk production.

Watch heat lamps carefully. They are mostly made very cheaply and are ready to fall apart at the drop of a hat. Be sure to maintain them so there are no frayed wires or loose sockets. Never hang them by the electric cord, but use a wire or chain on the loop attached for hanging. Heat lamps will warm lambs, but they also set fires, burn lambs and ewes, and can electrocute both sheep and shepherd. We used to have a black ewe who expressed her opinion of heat lamps in a way that allowed for no misunderstandings. Every time we put a heat lamp in the jug with her and her lambs she would eye it for a brief moment, then smash it to rubbish with her head. She knew.

RECORDS

Lambing is the time to start keeping records for each ewe and her lambs, if you haven't started already. The records will be used for a couple of important purposes. For one thing, they are necessary for evaluating the performance of the ewe so that she can be compared to her flock mates if your long-range goal is to keep only the most productive ewes and cull the rest. The records should include the identity of the sire as well, because he is half to credit (or blame, as the case may be). The records are also important to show both genetic heritage and individual performance of lambs to judge their suitability as replacements in your flock or to sell to others as breeding stock.

A ewe is valuable to the flock if she produces a lot of pounds of lambs in a given breeding-lambing cycle. Productivity means first of all that she must come into heat easily, settle on first breeding, remain healthy during gestation, and produce live lambs with little or no assistance from the shepherd. This paragon of ewes should then milk heavily to grow out her lambs rapidly.

The ewe's heat cycles are easily checked if crayons were used at breeding and should be part of the record. Her ease of lambing should also be recorded. A few difficult births are tolerable with a small flock, and one might even welcome some problems as training for a novice shepherd. However, a shepherd who is trying to lamb out three hundred ewes in two weeks is going to have little time to spare with tough births and should cull the ewes with problems.

Once the lambs are born, the ideal ewe should give lots of milk freely. She should accept her lambs eagerly and be an attentive, loving mother. Ewes who milk poorly or late, ignore their lambs, or even refuse to accept them or let them nurse, are not worth very much.

The real test of a ewe is how the lambs grow. Lambs should be weighed at birth, again in about three to four weeks, and again at some arbitrary time such as 60, 90, or 120 days; 90 days is a handy time to weigh and also give vaccinations.

Get all of the numbers together, go over them without emotion, and see which ewes are the best. Better yet, turn the decision making over to a hard-hearted computer so you can't let your likes and dislikes influence you. I predict that the cold facts will surprise you.

Put down dates of vaccinations of lambs, any treatments that were given, when ear tags were put in, tails docked, males castrated. Put down everything you can think of, because you never know when you might need the information.

Good records are priceless, so start them now and do them right. You'll be glad you did later for your own information. Also, you'll be pleased how impressed potential buyers of breeding stock are when they see complete records. Selling breeding stock really pays the bills.

EVALUATION

Lambing is a time when a lot of evaluation gets done naturally. A ewe who freshens with mastitis is a candidate for the truck, as are the poor mothers or the ones who are such good mothers that they kill other lambs to protect their own. This is also a period when you will be spending a lot of time looking at the sheep, so you will notice a lot of things that might slip by you in other parts of the cycle. No amount of advice from shepherds, veterinarians, magazines, or books can substitute for what the ewes will teach you if you just watch the flock and learn. If you spend a lot of time watching the ewes, one day you'll suddenly find to your pleasure that you can almost instinctively spot a ewe who is going to lamb in the next hour or so. The ewes give all sorts of indications, and it is just a matter of your learning to read the signs.

I know that it was great fun for me when I finally reached the stage in ewe watching where I was confident of my guesses at when they would deliver. I don't think it will ever cease to be pleasurable to be able to "read" a ewe, and recognize when she is ready, if she has any problems in store, and other signs that were a blank page to me for the first couple of years.

7

LACTATION

Aewe is in the business of producing wool and lambs, a job she does with remarkable efficiency. Her body is a factory that grows protein, fats, and bone on a year-round basis. Over the course of a year she will grow about ten pounds (up to twice that and more for some breeds) of essentially pure protein in the form of wool, probably the first fiber ever used by man for clothing and still the best in my unbiased opinion. This miracle fiber is coated with a soluble fat called lanolin that is a component of costly women's cosmetics and healing salves. One has only to shake hands with a sheep shearer after a day's work to find out what wonders lanolin can do for skin.

Still, for most growers, it is the production of lambs that is the ewe's major function in terms of cash value. This process is not spread over the year as is wool growth, but starts slowly after conception, then increases towards the end of gestation, and finally, after five months, produces lambs that generally weigh up to ten pounds or so each. Then the ewe's work really begins in earnest. While she has taken months to make ten-pound lambs, now she is going to give milk to those lambs at such a rate that some of them will increase in weight more than a pound a day, which means that some can weigh over a hundred pounds when they are weaned at ninety days of age.

This amazing ewe is going to eat hay and grain or some other simple feed and convert it into nourishing milk that not only grows lambs like crazy, but is the basis for some of the great cheeses of the world. She is a remarkable combination of feed-conversion factory, affectionate mother, and docile friend, making shepherding as rewarding as any occupation on earth.

NUTRITION: THE EWES

As one might expect, a ewe's nutritional needs are very high during lactation. These needs are so large that many ewes simply cannot eat enough feed to milk freely and still maintain their body weight, and a weight loss during lactation is the rule rather than the exception. A big milk producer that is suckling twins or triplets can be expected to come out of lactation with her ribs and hipbones pretty prominent.

The fact that a weight loss is to be expected should not be used as an excuse to withhold adequate feed. A lactating ewe with a single lamb will need the equivalent of about two pounds of shelled corn and four to four-and-a-half pounds of high-quality hay per day. If twins or triplets are being supplied, then even more feed should be given, as shown in the NRC tables in appendix 5.

A ewe can produce milk on lush pasture almost as well as she can on a grain-hay diet, but the shepherd should realize that many of our popular modern breeds of sheep are bred for farm flocks, and they will not do as well on grass alone as they will with some concentrate supplied. If good pasture is available, it should be utilized, but the shepherd will still find that some added grain will more than repay its cost with increased growth for lambs.

If hay is being fed during lactation, it should be the best hay of the year. As any dairy farmer will tell you, milk production is very sensitive to the quality of the hay fed, and top yield will come from the leafy, soft stuff with the highest protein and digestibility. You should have fed your poorest hay during early gestation and gradually improved your offering to reach a gourmet quality at this point.

If you simply don't have any excellent hay, you will still have to provide sufficient protein for milk manufacture. Let's face it, lots of us run out of hay from the previous year just about the time we need that good stuff for the lactators. In such an event, sit down with pencil and paper and the NRC tables and figure out what diet can be used. With lower quality hay plus grain, you will probably find that the total diet is short of protein. It can be added in the form of soybean meal or some other high-protein additive. The ewes may or may not need the extra protein for basic survival and health, but they surely will milk better in both quantity and quality. Don't try to save a few dollars on feed during lactation: your stinginess will be repaid with stunted lambs that will require far more feed later to reach market weight and finish. A lamb has its maximum growth potential during the early part of its life, and the goal of the shepherd should be to take advantage of the highly efficient feed conversion of young lambs. One way to do that is to feed the ewes well.

In addition to top-quality hay and adequate grain, the ewes will need plenty of fresh water, more than at any other time in their cycle. It is obvious that milk is mostly water, so the ewes need it for that reason alone. Equally important, the ewes are consuming a lot of dry feed, and they need water to keep the rumen contents suitably wet for the hard-working microorganisms that are doing the digestion job for their woolly host. Moreover, the ewe who is not getting enough water will limit her feed intake and will not produce milk as well.

Just providing the water is important, but encouraging the ewes to drink it is also beneficial. The easiest way to make ewes drink more water is to make them thirsty by giving them plenty of salt, just as a bartender sells a lot more beer by giving away free pretzels and popcorn. Salt can be made available free choice near the water and should be mixed half-and-half with dical to give additional calcium and phosphorus that is used in making milk. A sheep mineral mix can be used too, to provide the salt and dical plus trace minerals such as selenium and cobalt. If you wish, some plain salt (not trace mineral salt that contains copper) can be added to the grain at a rate of about 0.5 percent to 1 percent.

The lactating ewe's need for salt has been demonstrated by some fascinating studies by Australian veterinarian Derek Denton (*Hunger for Salt*, Springer-Verlag, 1982). Denton found that sheep have internal controls that keep the sodium content of their milk at a precisely constant level. The consequence of this is that if the sheep gets insufficient sodium in its feed, water, and supplements, it will first draw on body reserves, then, when those are gone, the ewe will produce less milk in order to keep the sodium level constant. Therefore, the shepherd must be very sure that lactating ewes have all the salt they want, because if they don't their milk production will decrease. It's basic: no salt, no milk—or at least greatly reduced milk, because feeds such as grasses and corn contain very low levels of sodium.

Many sheep raisers add monensin to ewe rations during lactation. It is added for two reasons. One is to suppress the growth of the internal parasites called coccidia, mainly to reduce the incidence of infection in the lambs, and some producers and veterinarians swear by it for this purpose alone. Monensin also affects the rumen microflora, causing an increase in the species that make the fatty acids, which are ultimately made into protein, at the expense of those that make fats. A typical rate of monensin in feed is about ten grams per ton. Do remember, as I have mentioned elsewhere, that monensin in feed is not approved for sheep, and is also highly toxic in even slightly excessive dosage.

NUTRITION: THE LAMBS

The feeding of young lambs is something no two shepherds do exactly alike. There is a sort of continuum from those who just let the lambs run with the mothers to nurse and eat whatever turns up, to those who manage the lambs more intensively by providing nutritious and palatable solid food along with the ewe's milk almost from the beginning.

The principal difficulty with letting the lambs tag along after their mothers as in the children's books is that this sylvan scene invites poor nutrition because of competition for food, principally from worms. Unless the ewe flock is totally worm free, and that is pretty much impossible, the lambs will quickly get infested, will grow poorly, and some may even die. However, if the ewes are treated with a wormer at lambing, and the pasture was clean, it will take some weeks for the lambs to become really wormy. At that time, worming the lambs is futile as long as they pasture with the ewes because the worms will just infest them again.

I cannot recommend that lambs be allowed to run with the ewes in intensively grazed pastures. If the stocking rates are low, grazing of lambs with ewes is more practical. This is especially true in range grazing where the flock is on the move, so that by the time oocysts hatch to larvae, the sheep and lambs have moved on to clean ground.

Creep Feeding

A solution to this problem is to provide a separate place for the lambs to eat so that they do not have the opportunity to ingest worm larvae. The lambs can be put in a pasture of their own, but this is easier said than done. What is more practical is to have a so-called creep area, a place where they can go to eat, but the ewes can't. They should have been preconditioned to a creep area by having been provided such a place adjacent to the community pens where they were moved soon after lambing.

One can arrange paddocks, yards, and the like so the lambs have access to the ewes for nursing, but don't go out to pasture with the ewes. While the ewes are on pasture, the lambs can eat in their creep area.

A creep area is a place that is fenced off by a partition that has vertical slots about eight or nine inches apart that allow lambs to squeeze through but exclude ewes. A horizontal space about a foot high will work too. The exact dimensions depend on the size of your sheep so you may have to do a little experimenting. The creep area should be dry, light, free from drafts, and generally pleasant, to make it as attractive as possible to the lambs.

Creep Feed

Creep feed, the name given to feed provided to lambs in the creep area, should be available in the creep at all times, and it should be fresh and palatable. Leafy hay is always of interest to lambs, so have some of that. For best growth, they should have a grain mixture of some sort. Finding out what lambs like to eat as early solid food is a matter of trying different things, but studies at the University of Illinois suggest some guidelines.

Soybean meal is a clear favorite of the Illinois lambs, perhaps reflecting a patriotic attitude, further supported by their high preference for corn, that state's other big cash crop. Some shepherds offer straight, undiluted soybean meal as the creep feed for very young lambs, and some lambs consume it eagerly. We have tried straight soybean meal, and found that our lambs were not especially fond of it, preferring coarsely milled corn. Experimentation is the answer, in my opinion, but a soybean meal, corn, bran, and sweetener mixture would be a good starting point. Any creep feed mixture can be improved nutritionally by adding 0.5 percent salt and 1 percent ground limestone, as well as a vitamin premix that provides 1,000 IU per pound of vitamin A and 200 IU per pound of vitamin D. If you are a believer in adding antibiotics to feed, chlortetracycline (Aureomycin) can be added at a rate of 50 mg/lb. (100

Table 2   Feed Consumed in a Two-week Period (in pounds)

Weeks	1–2	3–4	5–6	7–8	9–10
Oat groats	0.00	0.05	0.88	1.02	1.43
Whole oats	0.12	0.29	1.46	1.65	1.25
Shell corn	0.11	0.30	2.77	7.83	8.02
Alfalfa hay	0.23	0.41	1.43	1.12	1.05
Alfalfa pellets	0.00	0.14	1.35	4.06	1.98
Wheat bran	0.27	0.55	2.01	3.17	1.08
Soybean meal	0.81	1.70	6.94	11.01	10.63
Linseed meal	0.03	0.12	0.72	1.06	0.65
Linseed pellets	0.00	0.02	0.65	2.62	3.51
Sweet pellets	0.27	0.75	3.54	4.08	2.19

Source: "Young Lamb Nutrition and Management," Robert M. Jordan, University of Minnesota, 1975.

g/ton). If white-muscle disease has been a problem, consider adding selenium and vitamin E as well.

Feed companies offer creep feeds in pelleted form that provide convenience, at a price. These feeds are made of the same materials you can mix yourself or have mixed for you at a grain elevator. They will cost you about 30 to 40 percent more than homemade mixtures, with little or no improvement in lamb growth. Pelleted feeds are somewhat more palatable to lambs than ground ones, but if the grains are ground coarsely, the lambs will eat ground feeds almost as well as pellets. It is important with any kind of lamb's feed to be sure that it is fresh. Whether provided in troughs or in self-feeders, be sure to clean out the old feed each day or oftener and feed it to the ewes, giving the lambs fresh feed. Remember that it is in your interest to have the lambs eat as much as they will, so cater to their picky tastes.

You may want to add monensin to the lamb's feed. It is not approved as a feed additive for lambs, so your vet cannot prescribe it for off-label use. Studies conducted at the University of Saskatchewan by Dr. Glyn M. J. Horton showed that monensin improved lambs' performance in almost every respect. At a level of ten grams per ton of feed, lambs gained more weight than other groups receiving more or less monensin. Not only that, they made their greater daily weight gains on fewer pounds of feed for each pound of gain. In addition, the lambs grew as well on a 13 percent protein ration as they would have done on a 15 percent protein ration without the monensin. All of this from a feed additive that controls coccidiosis as well. One would hope that monensin will be formally approved for sheep someday, but that is unlikely. Adding both Aureomycin and monensin to feed appears to be irritating to the lamb's digestive tract, so avoid that combination.

Bottle Lambs

It is a rare sheep operation that doesn't generate some bottle lambs. A fortunate shepherd may be able to graft all the extras onto ewes with singles, but most of us aren't that lucky. Actually, having some bottle lambs is probably an indication of a good flock and good management, because it may mean that the ewes had a lot of triplets. For whatever reason, the lambs are valuable, and should be taken care of. If laziness or lack of time demand it, the lambs can be sold as bottle lambs. There are always a few people around who will pay for orphans and raise them as pets or for slaughter lambs. These people are willing to pay from nothing up to a premium price, so you might want to test the market with an ad in the paper or on a radio trading program that small-town stations commonly have.

Lambs with poor-milking mothers can be supplemented, but they will generally do better if pulled off mom altogether. Lambs on a bottle are fine if you like to do it, but a self-feeder as described in the last chapter is a lot less work, and lambs generally grow better, too. There is also the consideration that hand-fed lambs grow up to be tame pet sheep who aren't very afraid of you. If one of those lambs happens to grow up into a 300-pound ram, his lack of fear of you may be a real problem when he decides to get playful.

I have already mentioned that a lamb milk replacer is the best substitute, nutritionally at least, for ewe's milk. It or any other substitute, including cow or goat milk, is pretty expensive, and the goal should be to get the bottle lambs weaned to solid food as soon as possible. They should be weaned no later than six weeks of age, and researchers at Agriculture Canada's Animal Research Institute have found that weaning at three weeks is most cost effective. Get those lambs onto solid food soon, because every day they are on a milk replacer means you are losing money. Make every effort to give them the food they like and get them chewing and ruminating.

HANDLING

Much of the handling of lambs during lactation involves training them to get along without their mother. Most lambs will be very attached to their mothers, but the clever shepherd should be planning ahead for the day when the lamb will be permanently separated from his dam. One of the best ways to weaken the maternal bond a little bit is to give the lambs a pleasant place to get away from their mothers and into the company of their peers. Such a place is the creep area, although even if it is set up with feed, water, shade, soft bedding, and everything short of a color TV and a rack of comic books, some lambs simply will not voluntarily go to the creep, preferring the woolly familiarity of mother. To get those apron stringers used to the creep area, they must be forcibly put into it often enough to give them the idea that it is not such a bad place after all. You will find that lambs learn to eat grain by their mothers' sides very early in life, even at just a few days old. All you have to do is try to convince them that the grain in their creep area is just as good or better.

We use a creep area that adjoins the area where we feed the lactating ewes their daily grain ration. Both ewes and lambs enter the feed area through the same gate, but the ewes have to make a right turn to get to the grain troughs. The lambs' creep entrance is straight ahead, so in the mad rush many of the lambs simply fail to make the turn and head unswervingly into the creep area. A lot of the lambs turn with the ewes, of course, so while the ewes are oblivious to their surroundings in their intense concentration on eating, we go through their area and shoo the rest of the lambs into the creep. After a couple of weeks of this, most of the lambs are trained to go into the creep without urging. There are always a few who duck and dodge and have to be pointed in the right direction, but that keeps life interesting. Once all of the lambs are in the creep we leave them closed in there for a while so they can discover that it really is a lot more pleasant than out in the dry, dusty feedlot.

Another way to separate lambs and ewes is to make the ewes leap a low barrier that is too high for the lambs. A height of about seventeen to eighteen inches seems about right. Put feed out for the ewes and open a gate over the barrier. The ewes will sail over, leaving the lambs behind. The first two to three days, lambs will be in the way and get tossed around, but they quickly learn to stay back from the hurdle to avoid getting beat up by their mothers as they leave. This may sound bizarre, but it works beautifully. One year we had a badly overweight ewe. She was too fat to be able to leap out, so she didn't get her grain and slimmed down just fine.

Lambs need various treatments for diseases, of course, and the creep area makes a handy place to capture them. The creep area is a good place to spot lambs that are doing poorly. In the company of its fellows who are gamboling about or snoozing peacefully, a droopy, dejected-looking lamb stands out like a sore thumb. Catch those lambs, find out what is ailing them, and give them appropriate treatment. This opportunity to observe and treat lambs is so valuable that I favor a creep area even if the lambs are allowed to run with their mothers. A sick lamb can easily hide behind its mother until it finally is found dead. Get them away for even a few minutes, and the ones with troubles can be spotted easily. Many sheep raisers build portable creep areas—fences on skids—that can be towed around from pasture to pasture using a tractor. Some elaborate ones even have roofs to provide shade.

In order to utilize our pastures efficiently as well as to keep the lambs from acquiring worms, we send the ewes off to pasture while the lambs are confined to the creep. This not only makes maximum use of the pastures, but it gets the ewes used to being lambless for at least part of the day. In spring the scheme is especially useful—the ewes are eager for food and can be given a whack at the first, extra-tasty grass of the year without being given the chance to overgraze the tender new cover, because they are not allowed to be there too long. Later, when the shepherd wants them back, they are satiated with lush greens, and their bags will be tight with milk so that they want to get back with their lambs in the worst way. They return with eagerness and joy.

The only trouble with this method is the noise level. When the ewes and lambs get back together, they all call for one another at the same time as they mill about trying to get the right lambs attached to the right udders. The resulting cacophony is truly unbelievable and is comparable, I think, only to one of those riots that take place at soccer matches now and then. Things settle down quickly, however, and soon nothing's to be heard except the wet quiet of all the lambs nursing at the same time, interrupted only by the slap of a nose against a bag as a lamb urges more milk out with nudges and whacks.

Weigh the lambs one or more times during lactation to check on their progress. You may want the weight information for a variety of reasons, if only to make you objective. If you don't take the time to weigh all of the lambs, at least use your eyes and pick out the ones who are not gaining and growing along with the pack. Are they sick? Is there a triplet who is not competing for milk? Is a lamb wormy? Now is the time to find out.

MEDICAL: THE EWES

Mastitis

The bugaboo of lactation is mastitis. The shepherd must be alert for mastitis even before lambing and keep up the vigilance until after weaning. Early detection and treatment is essential because once milking capacity is lost, it never seems to return, and a ewe that cannot nurse a couple of lambs is virtually useless.

Mastitis is an inflammation of the udder caused by invasion and growth of one or more pathogenic bacteria or other infective microorganisms. In the early stages, the milk will be lumpy or have strings of material instead of being smooth. Also, a healthy udder will be soft, pliable, and elastic, without lumps or firm regions. If the udder is hard or lumpy, squirt some milk into your hand or a stripping cup (you can get these from dairy supply houses) to check it. Close to lambing, either before or after, a ewe's bag can be very firm and swollen, but the milk should look normal.

As mastitis becomes more acute the bag becomes lopsided as one side is milked out and the other is not. Lumps may be grossly visible even from a distance, and the ewe may walk painfully or limp, sometimes kicking at the udder with a hind leg. When things have gone this far, treatment is urgently needed. Remember that a ewe can come into lambing with a fully developed case of mastitis, so by lactation time things may be really out of hand.

Our veterinarian is a wise and sensible person who doesn't believe in discarding old methods just because they are old. His advice for treating mastitis is to use milking-out of the ewe as a primary treatment. Mastitis was treated this way for hundreds or thousands of years before sulfonamides and antibiotics were even dreamed of, and we should not abandon this harmless and simple treatment. The principle is elementary: milking removes excess fluids, pus, and many of the live microorganisms that are causing the problem in the first place. If milking is done frequently, the pathogens do not have as good an opportunity to increase in numbers and invade healthy tissue. Removal of the infective "bugs" helps to give the immune system a chance to get ahead of the infection, and only that will effect the final cure. Chemical treatments can slow down an infection, but the immune system can finish them off for good. Milking-out is also useful in preventing the walling off of pockets of infection that carry the potential of later reinfection should they rupture. Milk-out as often as you can take the time to do so. Frequent milking relieves pressure and pain, and the ewe is more likely to let lambs nurse to help with the job. Milking-out a ewe with mastitis is obviously painful to her, but rapid improvement after even one milking is the reward. The use of udder balms, warm towels, and gentle massage of the bag is worthwhile too. Anti-inflammatory drugs (even aspirin at one adult-human tablet per fifty pounds of sheep given every two to four hours) may help.

Chemical therapy is also called for in our experience, in addition to milking. Our veterinarian does not recommend the use of teat infusions if milking-out is being used as treatment. He thinks that teat infusions are for farmers who are too lazy to milk the animal.

Systemic treatment does seem to be effective. We give triple sulfa boluses orally for three days and simultaneously give injections of penicillin, which we continue as needed. I don't mean to imply that this is the treatment for your ewes, but it works with whatever organisms we are fighting. Get your vet's advice or talk to area sheep raisers to find out what works for your area. A veterinarian can commonly identify the pathogenic organism from the appearance of the milk, or culture the organisms and check their susceptibility to various antibiotics. Use all of the tools at your disposal, because loss of a bag is big trouble.

Prevention of the spread of mastitis should be attempted too. Isolate the affected ewe and lambs if you can because lambs who steal milk will spread the infection from ewe to ewe. When you milk her out, do so into a container and dispose of the infected milk elsewhere. When moving an infected ewe from a pen, lime heavily and bed well before using that pen again. There are vaccines that may help, but you will need to know what organisms you are dealing with, so get a culture of some milk from affected ewes. Be warned that treatment for mastitis is usually unrewarding, so don't be discouraged if you fail to stop it—you're not the first to fail to do so.

Lactic Acidosis

Acidosis can appear during lactation as it can at any time when grain is fed heavily. Treat such cases as described in Flushing.

Grass Tetany

Ewes on lush early pasture sometimes are affected by a condition called grass tetany. Happily it is not common, but it can happen anywhere, especially in April-May. Often ewes are just found dead, but before death they walk stiffly, hold their heads upward, twitch, and are wild-eyed. They go down, paddle with the legs, and rapidly die. Grass tetany is a result of magnesium deficiency, sometimes accompanied by calcium deficiency. To treat an affected ewe, give 50–100 ml subcute injections of 20 percent calcium borogluconate solution with added magnesium (often called Cal-Mag). In addition one can give subcute 50 ml of 25 percent magnesium sulfate in distilled water. The above injections should be given at several sites, because of the large volume of solution.

MEDICAL: THE LAMBS

Sore Mouth

Lambs can be vaccinated very early for sore mouth if it is a recurring problem on your farm. The ewes should have been vaccinated during gestation, long enough prior to lambing that the scabs have fallen off. If the form of sore mouth present on your place is the type that makes scabs on the lips of the lambs, you can still vaccinate the lambs during lactation. This type is not too serious in any case, and will only cause lambs to go off feed for a short period as the disease runs its course.

The main danger of common sore mouth is infection of the ewe's teats. An affected ewe will not let lambs suck, and she may get mastitis as a result. Be sure ewes are protected. With well-protected ewes, vaccination of lambs is usually not needed.

There is another strain of sore mouth that primarily infects the gums and throat instead of the lips. This strain spreads more quickly than the other variety, so early vaccination is a must, preferably in the jugs. I call this a strain of sore mouth because affected lambs will not "take" a sore-mouth vaccination, implying that the antigenic signature of the virus is the same as ordinary sore mouth. The gum type is not ordinary, however. We acquired it unwittingly by bringing an infected lamb onto our place. The lamb later came down with an active case that spread to lots of our lambs before we could vaccinate them. When it appeared, we were told by the breeder that it was nothing serious. It was serious enough to cause at least two lambs to die from starvation because they couldn't swallow. We should have been more careful. Protect your flock by not trusting anybody, even supposedly responsible breeders. Quarantine every lamb or sheep that comes into your flock.

We finally rid our flock of both kinds of sore mouth by revaccinating all ewes and by vaccinating all lambs in the jugs. One can take a chance that the flock will not be reinfected from buildings, equipment, or soil that might harbor the virus, and not revaccinate unless an outbreak occurs. Some veterinarians suggest this approach; others say to vaccinate annually. It does seem that once sore mouth has gone through the flock two or three times, either by vaccination or natural infection, it goes away to stay.

Enterotoxemia

Enterotoxemia (overeating disease) is the most important clinical problem caused by *Clostridium perfringens,* and it is generally not terribly threatening until passive immunity from colostrum wears off. The ewes must be regularly vaccinated against *Cl. perfringens* in order for the antibody levels to be high in their colostrum.

As mentioned before, early lamb scours are generally caused by these bacteria and can be halted by oral tetracyclines such as Terramycin. Another antibiotic, neomycin, is commonly added to milk-replacer formulas for prevention of scours.

Enterotoxemia from the type D strain or hemorrhagic enterotoxemia from the type C strain are both deadly to lambs. Because they are caused by excessive growth of *Cl. perfringens* from undigested carbohydrates that get into the intestine, the incidence is much higher among the lambs that eat the best. This means that the lambs that die are usually the fastest growing and healthiest looking ones.

Death is usually so rapid that the lambs are simply found dead. If lambs are observed with symptoms, they can sometimes be saved by prompt administration of antitoxin in large amounts, 10 to 20 cc per lamb. Affected lambs jump awkwardly with their hind legs, fall down and act uncoordinated, may make paddling motions with their legs, then get up again, or they may hold their heads against buildings or posts. These symptoms can be caused by other diseases that affect the nervous system, but enterotoxemia is a prime suspect in nursing lambs or those on feed. If the disease is not too far along, antitoxin will bring about prompt recovery. Take recovered lambs off feed for a few days.

We had a wether lamb one year who showed all the above symptoms and recovered when given antitoxin. Every time we tried to return him to grain feeding, he'd get the disease again. He had been vaccinated, but apparently his immune system was not making antibodies. We finally had to raise him out on hay alone.

If ewes were not vaccinated, lambs should be vaccinated at a week of age, and again two to three weeks later. According to studies at Cornell University, if the ewes were fully immune, lambs need not be vaccinated until three to four months of age. Ewes kept for replacements should be revaccinated a month later. Feedlot operators routinely give all lambs arriving in the lot at least one vaccination on arrival.

Some producers brag that they never vaccinate for anything and never have enterotoxemia or any other disease problems. This may be, but frankly I doubt it. It is true that if lambs are never given much high-energy feed such as grain, then the likelihood of diseases like enterotoxemia is very small. But the likelihood of getting the lambs to market before they are almost yearlings is pretty small too. If you prefer to keep lambs on limited feed, and that mostly hay, you probably can get by without vaccinating against enterotoxemia, but you had better plan on selling your lambs as feeders when they reach seventy or eighty pounds.

One final note: a food technologist at the USDA Western Regional Research Center in Albany, California, has determined that it is growth of *Cl. perfringens* in the intestines that causes the familiar human reaction from eating beans. Before you are tempted to immunize yourself or someone close to you, let me remind you that an antibody circulating in the lymphatic system still won't stop the gas in the intestine, for the same reason that passive immunity in lambs doesn't do much against baby lamb scours. I suppose you could sprinkle some neomycin crumbles over your next bowl of beans, but I don't think I'll try it.

Scours

Once the lambs are out of the jugs they have generally gotten past the stage of the early scours caused by *Cl. perfringens,* although Terramycin orally can be tried for scours during the first week or so.

After the lamb has passed the first tarry feces and then the bright yellow, sticky type, it will begin to nibble on solid food. Its droppings should begin to resemble miniature versions of the pellets or berries that older ewes produce. Any departure from this should be viewed with suspicion.

If a lamb has white diarrhea, the usual cause is excessive consumption of milk, either from an overproductive ewe or from an overeager shepherd. The cure is reduction of the milk supply. If the lamb is on a bottle, just reduce the feeding amounts. If the lamb is nursing a ewe, catch her and milk her out; this will usually cut the lamb down enough to effect a cure. To really do it right, mix up some of the electrolyte solution described in chapter 5 (1 pkg fruit pectin, 1 tsp Morton Lite Salt, 2 tsp baking soda, 1 10 ½-oz. can beef consommé, and warm water to make 2 quarts). Withhold milk and feed the electrolyte solution two to three times daily until scours cease, then continue for another twenty-four hours before returning to milk.

If the scours are brown or gray, then the cause can be any one of a number of things. Coccidiosis can be controlled by monensin in the creep feed if they are eating enough of it. Worminess can also cause diarrhea, so have some fecal samples checked. Either coccidia or worms are easy for your vet to identify in feces.

If the cause is some microorganism, that is harder to treat without knowing what organism is the culprit. It is helpful to check with neighbors, area veterinarians, and others who might know what the local causes of sheep scours usually are. Otherwise, it is a matter of you and a vet experimenting to find what antibiotic might control the problem. You need only look at a catalog of veterinary supplies and see how many scours remedies are offered to suspect that maybe none of them is any good or else there wouldn't be so many different kinds for sale.

You can never do harm and you might cure the scours by administering the electrolyte solution to prevent dehydration and by also giving some mechanical bulk such as Kaopectate or Endomagma to slow things down a bit. One school of thought believes in feeding nutritious mixtures of electrolytes, powdered milk, eggs, ground grain, and the like and avoiding medication altogether, and there is considerable clinical support for this idea.

If you practice hygiene and feed your sheep correctly and still have a scours problem year after year you should seek a long-term solution. Eliminate worms or coccidia as a possible cause, then work with a diagnostic lab, your vet, an extension veterinarian, or other experts until you find the answer.

Pneumonia

If the lambs survive scours, *Cl. perfringens,* and starvation, they may still have to face pneumonia. The shepherd should recognize at the outset that pneumonia is really a symptom and not a disease, in the sense that it cannot necessarily be traced to a single microorganism.

As I mentioned in chapter 6, vaccination with nasal PI-3 vaccine may be of some help in preventing infection by other organisms, though this is not proved. Adult sheep and lambs can also be vaccinated against bacteria called *Pasturella* and *E. coli* that are thought to be responsible for much pneumonia, and many veterinarians recommend that. Beyond those vaccines there is little that can be done by way of immunological protection. Pneumonia is brought on by a number of different organisms, and it can also be brought on by certain weather conditions. Cold, damp weather favors some outbreaks, while others seem to be triggered by dry, dusty conditions, and yet others appear in seemingly perfect weather.

Immediate treatment is essential in controlling pneumonia in lambs. At the first cough or sign of raspy breathing, give them help. Put the lamb up to your ear and listen. The breathing should sound quiet, not rough or rattling. You can feel the raspiness with your fingers or with a stethoscope if you know how to use one.

At the first signs of pneumonia, get out your syringe and needle. Only experience can guide you as to what antibiotic to use. We usually start with a penicillin, and then change to a tetracycline such as Terramycin if the penicillin produces no improvement. Some shepherds prefer tylosin (Tylan). Again, work with your vet, who can also prescribe off-label antibiotics.

As with scours, it is a matter of the individual producer's working with his or her veterinarian to recognize what pneumonia-causing organisms are present on a given farm, and learning to treat against them effectively. No matter how hard you try, some lambs will still die, especially when they are afflicted with what is called quick pneumonia, because of the rapidity of death after the first symptoms appear. To make yourself feel better, remember that there are some lambs who have defective immune systems from birth, and no amount of treatment can ever save them from an infection.

Kurt Wohlgemuth, then extension veterinarian at North Dakota State University, told a story of the time when he was a brand new vet working with an old, experienced one, treating a bunch of feedlot cattle. One day, discouraged with the high death losses, he confided his depression and lack of self confidence to his boss. According to Kurt, his boss put his feet up on his desk, puffed on his cigar a few times, and said, "Don't worry, Kurt, you can't kill them all." So when you've just hauled another dead lamb to the burning pile or lime pit, and you're feeling pretty incompetent, remember the old vet's words, and take heart.

Aspirin

Some of the lambs that look sick may just feel bad enough to lose their appetite, and their major problem might simply be lack of food. They will not be able to fight disease if they lack nutrition. What they need is rest, good food, plenty of liquids, and an aspirin. Yes, they need an aspirin just as you do when you feel crummy, or as your child does when it has a cold or the flu. The liquid that lambs need is fresh, warm ewe's milk, but they won't start drinking it if they don't feel good. Give a lamb a baby aspirin or half an adult aspirin. In many cases you'll see them brighten up appreciably and start to eat again. If they can be kept eating, that is half the battle. Get into the habit of carrying a bottle of children's aspirin in your pocket so you can pop a tablet into a droopy lamb when you see one. It really works. If you are nervous about putting a pill down a lamb, dissolve the aspirin in a little water and use the solution as a drench. You should use real aspirin and not the substitutes mostly used for children these days.

Necropsy

When all of your amateur efforts and even the professional treatment of your veterinarian fail to save a sick lamb, don't quit even after the lamb is dead. Use the dead one to help you save live ones. You can send the body off to a veterinary diagnostic laboratory for a complete necropsy including examination of all the tissues, culturing of the material to identify bacteria, and so on. If you are having a lot of necropsies done, the costs can begin to mount up, and you may be tempted to save the fee by just disposing of the dead lamb. Don't do that until you have cut it open yourself to see what you can figure out. If you have no confidence in your own diagnostic ability, ask your vet to let you watch the job done professionally. Most vets will be glad to show you how and give you instructions in basic sheep anatomy as they go along. If your vet won't show you, find one who will, or locate an experienced and competent sheep raiser to show you what to look for. If nothing else, do it yourself and simply use your eyes. After you have cut open a few lambs you will begin to have a feel for what is normal and what isn't. Once you begin to notice things, then you are ready to look up descriptions of lesions in books or describe details to your veterinarian over the phone that will help him or her to make an educated guess at a diagnosis. All you need is a sharp knife (a disposable scalpel is better) and possibly some pruning shears to cut through ribs. With the lamb lying on its right side, cut the skin under the left front leg and fold the leg over the back of the lamb. Then extend the cut back to the hip. Cut through the rib cartilage and then fold the left rib cage back, breaking the ribs where they join the spine. Now the belly cavity is exposed for checking.

A stillborn lamb will have dark uninflated lungs, and the kidneys will have yellow fat over them. There will also be a covering of yellow fat over the ribs and chest muscles.

With young lambs you should be able to diagnose some of the common causes of death pretty easily. With type-C enterotoxemia the intestine will be bright red from bleeding. If a lamb died of starvation, its stomach and intestine will be empty (A nursing lamb will have a stomach full of curdled milk.). Also there will be no fat around the kidneys or on the ribs and rib muscles, it having been used up for energy. The kidneys will be dark colored too. Some lambs die of impaction of the digestive tract by dirt or dry feed, and the intestines will be empty beyond the point of blockage.

A lamb that died of pneumonia will have dark, bruised-looking patches on the otherwise pale pink lungs. The dark parts will be toward the bottom front part of the lamb, with normal pink color to the top and rear of the lamb. Don't confuse that with dark parts of the lung that were down when the lamb died and blood drained into that part.

Now and then you'll find one that looks perfect to your untrained eye, but then remember that such is the case for many lambs that a trained veterinarian examines too. Keep in mind that a dead lamb is still making you money if it tells you something about lamb deaths.

Observations

Lactation is a time to make lots of observations, both of ewes and of lambs, but particularly of the lambs. You will have no trouble spotting scouring lambs, and one that is coughing is a candidate for pneumonia treatment. Keep your ear tuned for coughs, and if you hear one, find out which lamb did it, and treat it.

Starvation

A very important thing to be looking for is starving lambs. You might think that this sounds silly, but a survey of sick and dead lambs examined by the veterinary diagnostic lab at South Dakota State University in Brookings showed that the major cause of death of young lambs is starvation. If you are thinking that there is no excuse for that, I couldn't agree more.

The cure for starvation is food, and diagnosis is made by spending some time each day looking at the lambs, carefully. A starving lamb will be weak, humped up if standing, have sunken eyes and droopy ears, and just not look well in general. It will also not weigh very much. Pick one up and it will not feel hefty. Its stomach will not feel full to the touch. When you find a lamb like this, check its identifying number, and go find its mother. Check the ewe's bag to see that she is really producing milk and try to get the lamb to suck if she is. She may be making milk, but a more aggressive sibling may be getting all or most of it. If the lamb sucks, it will probably be all right, and may have just become separated from its mother, but it will bear watching for the next couple of days. If the lamb is pretty weak, it is wise to give it a meal of two or three ounces of milk replacer to get it going again. If the lamb refuses to suck, check for sore mouth since it may not be sucking because of pain. If it has sore mouth that badly, you should treat the scabs topically with injectable vitamin B12, and should plan to stomach feed it daily or more often until the sore mouth subsides to the point where the lamb can suck again.

You may find that the ewe has rejected the lamb. It is usually impossible to get a ewe to reaccept a rejected lamb. If she pulls away and won't let the lamb nurse but accepts it otherwise, the problem may be sharp teeth that hurt her teats. Use an emery board to file off sharp edges and corners; that usually solves the problem. If she still won't let it suck, you should probably take the lamb to the bottle jug. If it is a single lamb, make sure the ewe doesn't get mastitis from not being milked out.

Starvation can be prevented by frequent use of your eyes, so be watchful. Keep the aspirin and the stomach feeder handy, and use both when called for. A lamb saved is just like having an extra one born.

Slow Growers

Be alert for slow-growing lambs. They may be almost starving but not quite. Perhaps the ewe has a poor milk supply, or the lamb can't compete with a healthier sibling. Maybe they are wormy. Watch those slow ones, because they may be stunted for life if they don't get going when they are young.

Broken Legs

Lambs may get their young legs broken, usually by an aggressive ewe or even by their own oafish mother sitting on them. Happily, lambs' bones heal miraculously quickly. The idea is to immobilize the leg. Simple splinting with wooden strips such as tongue depressors and some tape is useful for routine breaks. For more serious breaks, use some of the foam used for pipe insulation to hold the leg together with bone ends in position for rebuilding. Wrap that in tape, then enclose in some PVC pipe split lengthwise, applied and taped, to make the whole assembly stiff. Be sure to make the PVC shorter than the foam, so it doesn't cut into the leg. The leg and the splint should share the weight to make the healing work. The leg should knit together in three to five weeks.

If a break is up in the hip or shoulder, fold the leg carefully, strapping gently with tape to keep it from moving around, and suspend the folded member off the ground with tape over the back of the lamb. That way the leg will not be used and will heal itself. For these maybe six weeks is needed, but usually less.

I once splinted a lamb's right front leg, but it didn't stay in place the way I intended. The leg healed perfectly except that the hoof pointed ninety degrees to the right. The lamb cared not a whit, and also could sure turn a right corner quickly.

Stiff Legs

Lambs with stiff hind legs should be investigated to see what the cause is. It may be that the animal is constipated. Watch to see if it tries to defecate. It may strain a bit if constipation is the cause. Give it a couple of tablespoons of mineral oil. If it is eating dirt, the constipation may be mechanical in origin. We even had a lamb die from eating the leftover ground limestone in feed troughs. To stop dirt eating, give lambs an injection of 1 or 2 cc of an iron dextran solution.

Similarly, stiffness can be a symptom of enterotoxemia, especially if accompanied by hopping and convulsions. Give *Cl. perfringens,* antitoxin, about 10 cc for a small lamb of less than twenty pounds or so and more proportionally for bigger ones, or as needed. Recovery will be dramatic if this is the cause and if the enterotoxemia was caught in time.

If the stiffness gives way to paralysis and the lamb drags its hindquarters along, the cause may be tetanus. Tetanus antitoxin will relieve the symptoms, though not as dramatically as does the *Cl. perfringens* antitoxin. Also treat the lamb with penicillin and keep administering antitoxin until symptoms disappear. If an infected wound is evident, open it up and clean it out with hydrogen-peroxide solution.

Finally, stiffness in the legs can be caused by white-muscle disease and can be corrected by injectable supplements of selenium and vitamin E.

The symptoms of all these conditions are similar, so do a little armchair detective work to supplement your observations and you'll be a better shepherd. Tetanus is only likely if you docked or castrated with rubber bands or a crushing tool. White-muscle disease is improbable if you have never had it in your lambs previously, unless you have changed feed or breeds in some major way. If you haven't seen lambs eating dirt, then dirt blockage is not very likely, and so forth. Instead of watching TV tonight, get out your sheep disease book and do some reading. You might save a lamb or two. Less common causes of stiffness are polyarthritis, and polioencephalomalacia (PEM).

Infections

Check docks and castration wounds for infections. The usual culprit is *Corynebacterium*, which produces a lot of pale green pus. Clean the wound, paint it well with iodine, and give an injection of penicillin.

Some lambs and ewes may also have abscesses from the same organisms. The abscesses will appear as lumps on the chin or face or, less commonly, on other parts of the body. These will feel hard at first, but as they progress, they will soften. When soft, they should be cut open with a scalpel, drained, cleaned, and swabbed with iodine. Use a Q-tip to get the iodine all around the inside of the cavity. Dispose of the pus away from the flock so as to reduce spreading of infections.

Bottoms

At the bottom of the list is a check of the bottoms. A lamb with a wad of feces stuck between its dock and its anus has big problems. It only takes a minute to clean it up, but if left alone it can cause terminal constipation. If that doesn't get the lamb, the mold and fly maggots will. If the lamb is dirty from scours, take a pair of hand shears and trim away all the dirty wool. The lamb will look better and feel a lot better.

Maggots can also get into damp parts of the wool, in sheep of any age, and cause serious problems, eating away at the sheep itself. Under heavy wool they can be hard to spot until lots of damage has been done. If maggots are found, clip off all the wool around the affected area, then clean the wounds with hydrogen peroxide solution, which will help sterilize the wound as well as float the maggots up and out. Be sure to check all over the affected sheep, as the maggots move around. Spray the sheep with an insect repellent to prevent reinfestation.

Check Your Bags

After a lambing time of lost sleep and frozen toes and fingers, followed by a lactation time of steady rain and knee-deep mud, you may be ready to check your bags at an airline or cruise boat ticket counter and forget it all for a while. But the bags I have in mind are the ones on your ewes. Detecting mastitis early is the only way to save a bag from loss, so do it.

The easiest way to check bags is to give your ewes their grain, then while they are absorbed in eating, move along behind them and check their bags with your hand. All but the jumpiest ones will tolerate the quick feel of a hand of a familiar shepherd. You'll soon learn what a healthy bag feels like. At the least sign of firmness or a lump, treat as described earlier. Unless that ewe is mighty valuable for her genes, she isn't worth a plug nickel if she can't nurse a lamb or two, so check those bags and save your ewes from a trip to market as culls.

EVALUATION

When you are watching the lambs, make mental notes or write them down if your memory is short. Lambs that are not doing well should be a hint that maybe a ewe is not doing her share. Inattentive mothers should be noted and possibly culled later. The lambs will be weighed late in lactation or at weaning, which will give you some hard facts to deal with, but don't discount the overall impression you get from simply watching. This is a good time to get some qualitative information to go with the quantitative data you'll gather later.

8

WEANING

eaning is a time of stress for all concerned. The emotional bond between a ewe and her lambs is quite strong, and a forced separation is cause for mental trauma of a high order. Both ewes and lambs make their unhappiness known by loud and persistent ba-a-a-ing, wild-eyed stares, and aimless rushing about. The noise level in the area reaches an annual peak at this time, attaining levels that would send an OSHA representative scurrying for a violation notice if it occurred in a public workplace. The shepherd who had the forethought to build the house and the sheep quarters well apart will be repaid with at least a modicum of quiet. This is also the time of year when many shepherds in suburban areas meet many of their neighbors for the first time—when the latter arrive on the doorstep to complain about the ruckus or send representatives from the local constabulary to do the same. To all but the most irate, an unyielding and calm explanation of the temporary nature of weaning sounds will usually restore neighborly relations.

The weaning cacophony can be minimized by some advance preparations, mostly by getting ewes and lambs accustomed to short-term separations before the final separation occurs. As mentioned in the last chapter, putting lambs into a creep area and sending ewes off to pasture each day is good training. In addition, the ewes' milk supplies can be reduced by feed and salt adjustments enough ahead of time to make the final cutoff less dramatic.

A really casual shepherd can let the ewes and lambs do the weaning themselves. At some point, many of the lambs will change over to solid food on their own, and gradually stop nursing. Alternatively, the ewe may decide that she has had it with nursing lambs and refuse to let them at the teats. A combination of these possibilities will accomplish weaning painlessly, even if a bit haphazardly. A few lambs and ewes will want to continue their relationship indefinitely, of course, but most will wean themselves naturally. Letting them do it themselves is probably all right for a small hobby flock but is not very practical for a large farm flock.

When to Wean

There is a full range of opinions among shepherds about when to wean, but there are some general principles that can be used as a guide.

Bottle lambs, including those on a self-feeding setup, should be weaned as soon as possible because they cost more in feed than they make in weight gains on a replacer diet. Some sheep raisers recommend weaning at three weeks, though this is quite early. Six weeks should be considered the longest nursing time for bottle lambs. Do everything to encourage bottle lambs to start on solid food so their rumens begin to develop. Once they are eating solid food regularly, wean them. They will drop in rate of gain and may even lose some weight, but in the long run early weaning is cost effective. Watch early-weaned lambs especially carefully to see that they make the change satisfactorily.

Lambs that are nursing can be left to suck longer. The only exception is if you want to rebreed the ewes quickly. Typical weaning times for ewe-sucking lambs are in the eight- to twelve-week range. We wean at about ten weeks.

Lambs gain very quickly on the combination of ewe's milk and a palatable creep feed or lush pasture, so don't be in too big a hurry to wean. What saves you money with bottle lambs does not make sense for lambs that are getting the real thing: straight from the ewe. Let the ewes do their excellent job of making lamb food of the highest quality.

Ample pasture and plenty of fresh water will provide all of a sheep's nutrional needs. These weaned lambs also get a free-choice grain mixture for quick growth.

NUTRITION

In order to get ready to wean, the feed to the ewes should be reduced over a period of a week or more prior to actual weaning. The simplest step is to reduce the amount of grain fed. If the ewes are being fed hay, the quality of the hay can also be drastically lowered or a reduced quantity can be fed. For ewes on pasture, reduce their grazing time. However it is done, milk production is easily slowed by feed adjustments.

As the date approaches, the shepherd should cut the milk supply further by reducing water as well as feed. Salt should be removed entirely, and don't forget salt added to feedstuffs. Feed can be sharply reduced over the last few days and water withheld altogether for at least the twenty-four hours prior to weaning; some say for longer. If mastitis has been a problem with your flock, consider putting sulfamethazine in the ewes' drinking water for three days, starting four days before the date of weaning, then withhold water the last day, as suggested by Dr. Charles Parker of the U.S. Sheep Experiment Station, Dubois, Idaho. If water is to be withheld very long, be sure also to withhold food so as to avoid impaction of the rumen with dry solids. This is not generally a problem, because most sheep will voluntarily slow down on eating if they lack water.

The lambs will of course need water to make up for the loss of ewe's milk. For bottle lambs, dilute the replacer more and more as weaning approaches, ending up with plain water. Have water available in tanks that are low enough for even the smallest ones to reach into. Some lambs will climb into tanks that are too low, so some compromise height has to be determined. Keeping tanks full usually discourages lambs from climbing in. If you can arrange it, the best way to provide water is with a demand valve that is licked or nudged to give a drink. Lambs adapt readily to these valves inasmuch as the transition from the ewe to a dispenser valve isn't too drastic. The small steel nipples made for pigs are perfect for lambs. You can mount three to four of them on a length of pipe fittings made up of T fittings and straight fittings, and attach the whole thing to a garden hose. The pipe with nipples is best mounted on a stand that can be moved, because lost water will make a muddy area. Lambs learn to use the nipples almost instantly. I was once asked by a visitor from New Zealand how long it took to train lambs to use the nipples as he was watching lambs suck vigorously. I told him that I had just turned on the water for the first time as he drove into the farm a few minutes previously.

Nothing is perfect, of course, and as you sit back feeling happy about keeping the water free of manure and dirt, realize that a demand valve is a perfectly wonderful way to spread sore mouth. If the lambs have been vaccinated, however, there is doubtless little problem, and the plusses outweigh the minuses.

The lambs will have been sampling creep feed for some time and there should be little fuss over the change to all solid food. But the composition of the feed should be changed to reflect the fact that they are no longer getting the protein from milk. You will want to increase the protein content by about 2 percent to make up for the missing milk. Be sure that the mixture also contains enough calcium to compensate for the loss of milk. We feed a 16.5 percent protein mixture after weaning. This ration contains monensin, which raises the effective protein content to about 18 percent.

Lambs enjoy hay and will eat some of it, but mostly they seem to like to select out favorite nibbles and waste the rest. My Scottish ancestry makes me rebel at wasted hay, and I have concluded that the best thing is to limit hay to lambs rather severely so they will eat all of what they are given. We have had

Lambs adapt readily to self-service waterers and get fresh water on call with little work from the shepherd.

very good results from feeding free-choice grain mix plus about a quarter of a pound of hay a day per lamb. You will have to see how much they will eat without waste, and you may want to feed a bit more hay, but they seem to prosper on a rather small amount. You'll find that you have to give some hay or in their quest for roughage they will try to eat the barn, fence posts, feeders, twine, your pants, or anything else that's handy. Try to strike a balance between severely limiting hay and giving them so much that they waste a lot.

The question of limiting hay sometimes brings strong negative reactions from sheep raisers. I can only cite my own experiences and those of a lot of modern shepherds who are not believers in a lot of hay in a lamb's diet. One year we fed a bunch of wethers straight hay for two months trying to get them to market weight. At the end of the sixty days we weighed them, and on average they had not gained a pound. The cost of that hay was a total loss. You can't convince me that feeding hay is a profitable way to run a sheep business, at least not a lamb-growing business. I have had old-timers tell me that they raised lambs on hay, and perhaps they did, but it seems like a poor plan to me. Pound for pound, corn is roughly twice as nutritious as hay, and it usually doesn't cost twice as much. More important, a grain mixture doesn't fill the lambs up as quickly, so they will get more feed into themselves and grow faster.

There is also the question of the relative cost of hay and grain when the lambs waste a lot of hay. If you can imagine lambs wasting two-thirds of their hay—which they will be most happy to do if you give them too much—and then combine that with the fact that hay is about 50 percent digestible, you will find that hay may actually be much more expensive than grain as a feed, in terms of what gets into the lamb. Happily, there is a solution that my kilted ancestors didn't have, and that is pelleted alfalfa. The cost of pelleted alfalfa is higher than plain hay, of course. However, it is sold with a guaranteed protein content, usually 15 to 18 percent, and the lambs will eat all of it, with zero waste. Pellets can be fed free choice, mixed with a grain mixture, or given out in measured amounts, whichever suits your feeding program. Some people feed pellets plus hay as the total ration, although that does not really provide enough energy for best growth. Some plain hay should be fed, even when you are using pellets, but it can be minimal—like one thirty- to forty-pound bale for every hundred lambs.

Commercial feeds often have added phosphorus, which should be avoided like the plague. Grain is rich in phosphorus and supplies a sheep's needs ordinarily. A diet with excessive phosphorus in relation to calcium causes water belly or urinary calculi in ram and wether lambs, in which mineral secretions, or stones, plug up the urethra.

If lambs are raised on a diet rich in grain, the ratio of calcium to phosphorus will be distorted from the optimal in favor of phosphorus. Addition of ground limestone to the feed to bring the ratio of calcium to phosphorus to about two to one or three to one will balance these minerals for a sheep's needs.

Addition of phosphorus to rations is something that the feed companies like to push very hard. I recall a feed company employee looking in a feeder at a ration of ours that was about 90 percent corn and asking whether I was sure that the lambs were getting enough phosphorus. I didn't know how to answer a question like that from a feed company man who had a Ph.D. in animal nutrition. I have no objection at all to feed companies making a profit, but I do object to them trying to shove extra phosphorus into my male lambs.

A lamb requires roughly three grams of phosphorus a day, pretty much regardless of its weight. If hay alone is fed, the lamb would have to eat somewhere in the range of two and a half to five pounds a day to get the three grams. On a straight corn diet, less than two pounds a day would provide sufficient phosphorus. The hay diet would provide calcium and phosphorus in a ratio of about 1.5:1 (Ca:P) to over 6:1 (Ca:P) depending on the type of hay. Enough calcium would be provided by about three pounds of grass hays or by less than a pound of typical alfalfa hays. In contrast, a diet of straight shelled corn provides very little calcium, and the hapless lamb would have to eat a truly phenomenal fifty-five pounds of corn a day to get enough calcium. Not that anyone would feed straight corn, but that gives you some idea of the amount of calcium in America's number-one feed grain. If that feed rep had asked whether the lambs were getting enough calcium, the question might have made sense.

What about adding phosphorus to a lamb's feed? On a purely hay or alfalfa-pellet diet, some additional phosphorus might be called for. The lamb would have to eat about five pounds of hay to get enough, and that is a lot of hay to jam into one lamb every day.

A far more sensible approach is to give the lambs a varied diet composed of hay, grains, oilseeds, and other foods that naturally contain balanced amounts of calcium and phosphorus, among other nutrients. The simplest feeding scheme is to feed a grain mixture that has enough finely ground limestone added to balance the relative excess of phosphorus. Hay can then be given in any amount preferred to provide a total diet, because hay has about the right ratio of Ca to P for sheep. If urinary calculi are a problem with such a diet, have your feed analyzed to find the Ca:P ratio rather than using average values from tables. If urinary calculi are still a problem, up to 2 percent ammonium chloride may be added to grain mixtures as a preventive measure.

ENVIRONMENT

Weaning is another time when the tightness of your fences will be tested. When the ewes and lambs are separated, they will try like the dickens to get back together. Lambs will find a crack to squeeze through that they have ignored previously, and ewes will leap fences with an agility and grace never seen the rest of the year. I recall being astonished one year to see a grossly overweight Montadale ewe sail over two fences like a deer to get back to her youngster. If your interior fencing keeps ewes and lambs apart at weaning, then you have pretty good fences.

I have previously mentioned that woven wire with the stays twelve inches apart is the preferred stuff for interior fencing because the ewes can get their heads unstuck if they push them through to eat on the other side. The one exception to this is the area where lambs are confined. Here, six-inch stays are required, or some altogether different sort of fencing. For places like creep areas, galvanized steel hog panels are handy. They are tight enough to confine lambs, but are short enough to let the shepherd step over them easily. The panels are relatively expensive, but creep areas are small so the total cost is not prohibitive.

It is also convenient to have one or more small pastures with lamb-tight fencing to let the weaned lambs use. This gives them an area to play around in but keeps them from rejoining their moms. Once they have gotten over the shock of weaning they can be run out to other pastures with less tight fencing. Many shepherds let the lambs feed ahead of the ewes in a rotation scheme to give them first crack at grass, although for most of us weaning is at a time of year when pastures are not at a peak productivity and keeping the lambs in a well-fenced, familiar feedlot area with hay and grain available is probably the best plan.

HANDLING: EWES

Don't let yourself be tempted to separate the ewes from the lambs and then forget about the ewes for the rest of the time until breeding season approaches. After all of the attention paid to detail during and after lambing, it would be kind of nice to goof off, but weaning is a time when that old enemy mastitis can sneak in on your flock. Even with feed, salt, and water reduced, the ewes will continue to produce milk, some of them more than others. The only way to monitor the ewe's progress is to palpate (feel) bags a few times. Confine the ewes in a small pen and quickly check the bags to make sure they are drying up as they should. Generally you can just look at the bags and palpate any that seem swollen.

It is a bit of a Catch-22 situation in that pressure of milk in the bag causes milk secretion to slow down and stop, but at the same time, a too-full bag

Lambs enjoy access to some hills and trees for a bit of recreation.

needs to have the pressure relieved by a little milking out. It is strictly a matter of judgment and you will have to learn by trial and error when to milk a little out and when to leave a tight bag alone. Some producers turn the flock out after weaning and hope for the best. If the flock is large in relation to the time and labor available for checking bags, this may be the only solution. If this is done, one must accept the fact that some bags will probably be lost, and you won't even know about it until the next lambing. You will lose the ewe, plus the cost of the feed that went into her over the intervening months. It doesn't take very many lost breeding ewes to justify the cost of checking their bags two or three times after weaning.

Handling Lambs

Weighing

The lambs will require little handling at weaning except to weigh them soon afterward. Don't pass up this weighing chore just because you are busy. Lamb weights are the basic data you need to evaluate your breeding stock, and this is the time to get the numbers, so do it. I also encourage you to buy a good scale to weigh both lambs and sheep, and to put it in a chute/corral/pen arrangement that makes weighing easy to do. Scales designed for pigs work great.

Shearing

Once the stress of weaning has subsided, you might consider shearing some or all of the lambs. There is no simple answer as to whether you should shear lambs. One consideration is the cost of shearing relative to the value of the wool from the lamb. If shearing a lamb costs you a dollar, you'd have to get a dollar's worth of wool from the lamb just to come out even. For the value of the wool, check local market conditions unless you sell to a niche market such as handspinners.

To complicate matters, recognize that you are paid for the weight of the wool attached to the lamb when you sell it. So if lamb prices are higher than wool prices, you might want to leave it on. However, the price for unshorn lambs is usually lower than for shorn ones with a pelt about one inch in length (a so-called number-one pelt). The key factor here is the value of lamb pelts, and that ranges widely from year to year—from zero difference to as much as $6.00 difference in the past decades.

There are also other considerations. If you are selling breeding stock, you probably will want to shear the lambs to make them have a more attractive appearance to prospective buyers. This is especially important with Suffolk or Suffolk-cross lambs because they are born with a dark wool that grows out white, or at least whiter, so shearing is needed to give the characteristic white and black pattern. Even with all-white breeds, a nicely shorn lamb will usually command a higher price—unless, of course, you are raising long-wooled breeds.

If lambs are being fattened in summer, shearing may be a worthwhile management practice to keep them cool and growing in the hottest parts of the season. Whether shorn lambs actually do grow better in hot weather is a matter of opinion, and you will have to decide for yourself, because the climate and heat tolerance of the sheep will be different for almost every farm.

If you choose not to shear, you will have to be alert for accumulations of dirt and dung around the rear of the lambs. Any dung tags hanging from the lambs can cost you dearly when it comes time to sell them. A buyer who sees dung tags will take them as a sign of poor management, just as he would with undocked tails, so the price you receive will be reduced accordingly. You may get stung twice for your inattention to detail. You will get a lower basic price because of the bad impression your lambs give, and the weight of the lambs may be arbitrarily reduced by some pounds for the weight of those hanging clumps. So, if you don't shear, at least take the time to clip off the tags around the rears of those that need this treatment.

If you do decide to shear, do it when the weather is favorable, which may not be easy to plan if you depend on someone else to shear for you. If you can, try to shear just before warm, dry weather because the shearing stresses the lambs, and then if they are out in a cold rain or a late snowstorm, the survival rate can be low.

In warm weather, protect the freshly shorn lambs from sunburn and from flies. Flies will literally chew holes into the backs of freshly shorn lambs, and mosquitoes can suck their blood. Protect lambs with insect repellent sprays. Treat any shearing cuts, and spray them with repellent to prevent flies from laying eggs in the wounds.

If you notice any lice or keds on the lambs when you shear, treat the lambs with a powder or spray, although most of the parasites and their eggs will go with the wool. External parasites will slow the growth of a lamb to nothing if they are left untreated. A shorn animal is very easy to treat, so don't pass up the chance.

MEDICAL

Vaccination and Worming

A second PI-3 vaccination is favored by some at this time. If the lambs are to be placed in a drylot, worming might be appropriate if they have been with the ewes in a pasture up to this time. You should wait a few days after shearing to worm them, because the combined stress of shearing, vaccination, and worming added to weaning may be a bit much for some of the little ones.

Rectal Prolapse

A medical problem that may appear about this time is rectal prolapse. In a rectal prolapse, a piece of the rectum is pushed out, turning inside out as it appears. A piece a few inches long may hang out, and it will get dirty, sunburned, dried, cut, and eventually split open if left untreated, resulting in the death of the lamb. I feel compelled to mention that the only time I ever had this problem with lambs was with lambs from one Suffolk sire, the original owner of whom also had prolapse problems in his flock, so it could be a genetic thing.

One treatment is to push the washed mass back inside and suture up the anus to prevent its coming out again. The trouble with this method is that if the opening is left loose enough for the lamb to defecate, the rectum will push out again. If it is tighter, the rectum may push against the stitches, swell, and plug up.

Another method that is widely used by both veterinarians and producers is to use a so-called rectal ring. A rectal ring isn't a ring at all, but a tube made of plastic pipe about two inches long and about an inch in diameter. There is a groove around the center of the pipe. The everted part of the rectum is left sticking out and the tube slipped inside the protruding rectum until the groove is flush with the anus. Then an Elastrator band is placed round the everted tissue, at the anus, so that the prolapsed portion of the rectum is pinched off by the rubber band which squeezes into the groove in the rectal ring. The blood supply to the prolapsed part of the rectum is cut to almost nil, so the tissue dies and eventually sloughs off, usually after a few days. Meanwhile the pressure of the rubber band causes the tissues against the anus to grow together so that the remaining rectum is grown to the tissue at the anus. The prolapsed part is effectively amputated, and the lamb is back to normal. When using a rectal ring, be sure that the tube doesn't get plugged up with feces. Clean the tube if needed and give a dose of mineral oil (two tablespoons) to keep things moving.

A more drastic method is used by some shepherds, and I have tried it with success. This is essentially the rectal ring method with no rectal ring. A piece of umbilical tape can be tied around the prolapsed tissue to help to hold onto the slippery mass. Then an Elastrator band is slipped over the prolapsed rectal tissue and placed snugly against the lamb's body. The band cuts off the circulation to the prolapsed tissue, and it dies and sloughs off in three to four days. The obvious difference between this and the rectal ring method is that the lamb cannot defecate at all. If the tissue does not slough off in four days, it can be cut off and the band removed. Needless to say, do not cut the tissues on the lamb's side of the Elastrator band, but on the other or distal side.

When the band is removed—and be aware that it may come off all by itself when the tissue is cut off—stand aside because feces may come out of the opening with almost explosive force, as you might imagine.

Lambs usually limit their food intake voluntarily when plugged up like this, but watch them closely. If they continue to eat very much, pen them away from food for a few days until the amputation is complete. Watch them for bloat and any other side effects.

I won't recommend this latter method to you, and I'm sure your veterinarian wouldn't either. Even a vet whose partner had badly botched a rectal-ring job for us didn't approve of a no-ring method. All I will say is that I have tried both the ring and no-ring ways many times and I prefer the no-ring way. Use your own judgment, and take your chances. Remember to protect the lamb against tetanus by giving 100 units of antitoxin when an Elastrator band is used, if you have had previous tetanus problems, and if the lamb did not receive vaccine previously. The best solution is not to dock too short and to get rid of any rams who sire prolapse-prone lambs.

OBSERVATIONS

As the lambs are weighed, look over the best ewe lambs as potential replacements. At the time of weaning, the qualities you see in the lambs may largely reflect their mother's traits rather than their own in the sense that you are seeing the results of heredity plus milking ability of the ewe. Make a preliminary selection now, and keep an eye on them to see how they progress on their own. You may want to separate these lambs and feed them for a little slower growth (more hay and less grain) if they are to become permanent members of the flock or are to be sold as breeders.

At the other end of the scale, be on the lookout for lambs that look small and weak, because they may have a very negative response to weaning. Some will not make the transition to solid food at all well, and these will bear watching. Some will eat the solid food just fine but will not drink enough water and will die of impaction. In studies at the U.S. Meat Animal Research Center in Clay Center, Nebraska, it was found that this condition is a particularly serious difficulty with lambs weaned very young (about ten days).

Evaluation

Weaning is a good time to consider strategies for marketing your lambs. If you want to catch an early market, feed the heck out of them. Alternately, you may opt to slow-grow them on pasture in order to finish them later for a higher price in the winter. Maybe you will decide to fatten them for a while and then sell them as feeders. You may decide on a combination of all of these, if that is appropriate. Any lambs that are to be kept as breeders should be taken away at some point and given more hay and less grain to keep their rumens in good shape. A low-roughage diet leads to a stomach condition called parakeratosis that is not desirable if an animal is to be kept beyond a year or so. Therefore, don't try to raise your replacements on ground or pelleted feed without including some long hay (not pellets or ground hay) or pasture.

Weaning is also the time to evaluate ewes. The weaning weights are in and that is your primary information. The ewe that gives the most pounds of healthy lambs at weaning is the best one. A ewe who gives runty twins is not nearly as valuable as a ewe who gives a big, growthy single that outweighs both of the twins put together. If you use computer software to analyze your flock, you will certainly need both birth and weaning weights.

Conclusion

Apart from the appendixes that follow, our exploration of the cycle of the ewe is complete. You can lay this book down now, having managed to wade through a lot of material—enjoyably I hope. And although you may be finished, your ewes are not. They are already back to Building and Rebuilding even as you read this. They're getting ready to make lambs grow wool by stocking up on food and resting up. May they prosper, bless their hearts.

APPENDIXES

APPENDIX I

Some Typical Sheep Facts

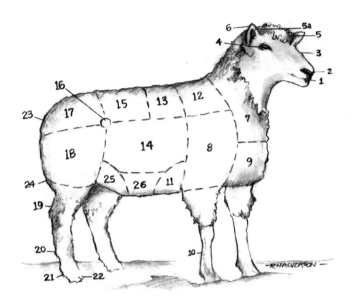

SHEEP PARTS

1. mouth
2. nostril
3. face
4. eye
5. forehead and forelock
5a. poll
6. ear
7. neck
8. shoulder
9. brisket
10. foreleg, including shank
11. foreflank
12. top of shoulders
13. back
14. paunch
15. loin
16. point of hip
17. rump
18. thigh
19. rear leg
20. pastern
21. dewclaw
22. foot
23. dock
24. twist
25. rear flank
26. belly

WEIGHT OF ADULT EWE: 120–325 pounds

WEIGHT OF ADULT RAM: 160–400 pounds

RATE OF WEIGHT GAIN OF LAMBS: up to about 1.5 pounds a day

WOOL CLIP: 8 pounds avg. (4–35) per year

PERIOD BETWEEN HEAT CYCLES: 16.5 (14–20) days avg.

DURATION OF HEAT: 24–48 hours

DURATION OF GESTATION: mutton breeds 144–147 days, wool breeds 148–151 days,
Finn & crosses 140–142 days

WATER INTAKE: adults, 1.0 gal/day; lactating ewe, 1.5 gal/day; lamb, 0.5 gal/day

RECTAL TEMPERATURE: 102–104° F

HEART RATE: 75 (60–120) beats/minute avg.

RESPIRATORY RATE: 19 breaths/minute

DAILY URINE VOLUME: 10–40 cc/kg body weight

DAILY FECES WEIGHT: 1–3 kg/day

PRINCIPAL USE OF SHEEP PELTS: covering sheep

USUAL SOUND: baa

APPENDIX 2

Marketing

Unless a few lambs are being raised as pets or for home consumption, there comes a time when they must be sold. That can be a pleasant, profitable experience, or it can be an event when you go to a lot of work only to be left feeling that you've been cheated or at least taken advantage of. The same is true for selling breeding stock and disposing of older ewes and rams. Finally, unless you are a handspinner, or love to make wool-filled quilts, you will have to try to sell the wool from your sheep. All of this is marketing.

The time to consider marketing alternatives is not a few days before you are ready to sell something, but long before you decide to raise sheep as a business, even if only part-time or as a hobby. Nobody should consider raising sheep if there is no convenient market for the products that will come from the enterprise. There are two sorts of markets: the ones that are already in place, and the ones that the producer develops from scratch. Unless you are the sort of person who thrives on the challenge of making something from nothing, you had better be sure there is a handy way of selling your sheep and wool.

LAMBS

Lambs are the main product of any sheep operation, no matter how big or small. Their ultimate destination is usually someone's plate, at home or in a restaurant, but their route there is very roundabout in some instances. How, then, do you get them to their destination? (Let's assume the lambs are in a proper state of fattiness or condition for slaughter, and that they are at a desired market weight, generally in the 105–135 pounds range.)

Direct Sales to Consumers

Many small producers, and some that are not so small, take a do-it-yourself approach and try to sell lambs direct to the consumer. To be a success, there have to be enough customers to buy the lambs. I realize this statement is so obvious as to be almost insulting, but there are many direct-to-customer marketing schemes that failed because this point was ignored. You will need people who are used to eating lamb, which greatly reduces your chances in rural areas, particularly in the Southeast and much of the Midwest. You also need people who are used to buying foods in quantity, because you want to sell whole lambs, not retail cuts. Potential customers used to be mostly affluent middle-class people, most of whom lived in city suburbs. That profile has changed dramatically in the past decades as more immigrants from North Africa and the Middle East have entered the United States. The Muslims among those potential customers prefer uncastrated male lambs weighing 50–80 pounds live weight. They often also want to slaughter the lambs themselves to meet religious requirements for halal slaughter.

No matter how many ideal customers are in your imagined sales area, you will still have to reach them. Word of mouth is by far the most effective means in the long run, and this means knowing a few people who can spread the news at work, school, at their church or mosque, club, or other social gathering place. Classified ads in a local advertiser or shopper newspaper can work too if the ads are really read by your potential customers. There are many Muslim students at universities who long for fresh lamb, and they can be your contact with a larger lamb-eating community.

Once you have reached your customers you will still have to get the meat into their freezers. In most regions you will find that it is far too complicated to try to sell packaged cuts or even whole sides because of health laws and regulations that govern slaughter and butchering, especially slaughter. You will doubtless find that it is illegal for you to sell a dead lamb, no matter where it was slaughtered, much less if it was done in your barn or under a tree. The way to do it legally anywhere is to sell the customer a live lamb, and then let them make the arrangements for slaughter and butchering. Some customers prefer to pay a locker plant to do all of the work, providing a box of neatly wrapped and labeled cuts made to the customer's specifications with minimum fuss. You have sold a live lamb, which is legal, and the customer has arranged for slaughter and butchering for personal use, which is also legal. The Muslim customers who want to do the slaughtering themselves can be provided a place at your farm if your state laws allow it. Check with your state's health department.

As to pricing, you can use local market prices as published in newspapers as a guide. If you think that your lambs are better in some ways than the run of the lot sold through conventional marketing channels, then say so and expect a higher price, but don't get greedy. If you can show your customers a clean farm with happy-looking sheep (although I'm not sure how one tells whether a sheep is happy or not), or can show that all the feed you give is natural or organically grown, or that they are raised on a special sort of feed (seaweed, herbal pastures, or whatever gimmick is unique to your lambs), then maybe you can get more for your product. If you have no local prices to use as a guide, then you should realize that a lamb carcass is about half the weight of a live lamb and that about 70 percent of that weight is retail cuts. Check prices at a local market, or ask a butcher what the current price of lamb carcasses is. Some producers set a fixed price per lamb and let the buyer choose which one they want. That certainly eliminates problems of weighing. Otherwise, the live weight can be taken at your farm or at the locker plant.

Your price should be on the farm, or else should include a charge for delivering the live lamb to the locker plant of your customer's choice. You will probably end up delivering many of the lambs sold. Be sure to make all of the costs clear to the customer before a deal is concluded, and remember that the customer is always right. Do make sure that they realize that the yield in cuts will be only about a third of the live weight, even less if any cuts are boned. The last is important if the customer is used to beef, which has higher yields.

Local Auction Sales

A common way to sell small numbers of lambs in rural areas is to take them to a local livestock auction. Many people come to those sales to buy animals for the table, and you will have little trouble selling your lambs. You also probably will have very little trouble fitting the money in your wallet, because the price you get will generally be well below market price. Most of the people who come to buy at livestock auction sales are either bargain hunters for their own table, or they are people who make their living by buying animals at local auction sales and reselling them to a packer or at a terminal market. Naturally they cannot pay market price, or they would have no margin of profit for themselves. Most local markets put a maximum number of middlemen between you and the ultimate consumer, and the price you get shows it. Also, you will do all of the hauling and pay a commission to the auction yard. For the most part, the buyer at a local auction gets a good deal, and you don't. There are notable exceptions to this generality, so check locally. Some local auctions have sweetheart deals with packers, or can carry transportation costs on the backs of some other enterprise, or have various ways to pay a high price for some types of animals that they want badly. Don't discount the area auction without checking, but don't expect to find good deals very commonly. Be suspicious.

Terminal Markets

Probably the most common method of lamb marketing for the average producer is to get the lambs to a large terminal stockyard, where they are purchased by buyers for the big packing plants. The prices will generally be higher than at local auctions, and you might even get to keep some of the money. What I mean by that is that there are a lot of costs involved. First, there is the cost of getting the lambs to the marketplace. If you do that yourself, you know the costs. If you hire a trucker to haul your lambs, you will pay a going rate plus insurance. The trucker will also have a pickup charge for each load and possibly other charges. You will consign your lambs to a representative at the terminal market. Usually a marketplace of any size will have a number of competing organizations that are in the livestock selling and buying business. They get a commission for their services, and for that reason the person who does the selling for you is often referred to as your commission man.

In addition to the charge for selling, you will find yourself laying out money for hay, health inspections, use of the yard facilities, and other costs. Each of these costs is small, but they add up and will generally total from as little as two or three dollars per head to much higher figures. In an informal survey of our local sheep raisers' association, the total of trucking plus the commission and other costs came to between five and nine dollars per head for the market at South Saint Paul, Minnesota, which is about 150 miles from the farm area. Compare expenses and decide whether a large terminal stockyard is for you.

The key person in the large terminal market is the commission man. First of all, you want to have a person represent you who is savvy about lambs and sheep, so try to find a firm that has one or more persons who deal exclusively with sheep rather than one that has a pig or cattle expert who also handles a few lambs once in a while. If you have a choice of firms, ask around your area to find out which one has a reputation for getting top prices and paying promptly.

Once you have decided on a firm to represent you, get a load of lambs together and send them off to see how they do. One way is to select some that look like they are big enough and ship them. Doing this will generally not get you the best possible price though. To get a top price, the loads you ship should be reasonably uniform, and they should be aimed at a particular market. For example, if you are selling lambs as feeders, then they all should be ewes or wethers (no rams!) of about the same weight. They should have no tails or dungy rear ends, and should be reasonably clean. This doesn't mean that you have to run them through the car wash before you send them, but don't send any really dirty lambs or the buyer will be suspicious that you are a poor manager, and that there may be other, hidden things wrong with them. Spiffy-looking lambs get a better price even if they are actually no better than cruddy-looking ones. Whether feeders are shorn doesn't usually make a great deal of difference, but a freshly shorn feeder saves the feedlot owner the bother and expense of doing it, so you might get a bit more money.

If you market lambs as slaughter animals, you can make yourself a few extra dollars by paying attention to certain details. First of all, be sure that all of the lambs will grade at least Choice, or else they will have to be sold as feeders regardless of their weight. (I'll discuss grading lambs a little farther on.) Next, make every effort to have the lot as uniform as possible; in other words, all lambs should be about the same weight, the same general breeding (blackfaced or whitefaced), all either shorn with number-one and -two pelts or all unshorn, and so forth. The more uniform a lot looks, the more favorably the buyer will react to it. Uniformity tells the buyer that the shipper is doing a good job of raising and managing sheep, and that is worth money. A given lot can be made up of all ewes and wethers or can include a few ram lambs. If the ram lambs make up only 10 or 15 percent of the lot, there will generally be no effect on price. If you have a large number of heavy (125 to 140 pounds and up) ram lambs, they should be grouped separately or else they will drag down the price of the whole lot. One way to arrange the sale is to take a dock on the rams so that only they are affected rather than the entire lot.

Details vary from place to place, year to year, and with time of year, but heavy lambs can be subject to price docking. This is usually not the case in spring when heavy lambs are in short supply, but later in the year they become abundant, and penalties are exacted. Usually a top weight is established by a packer or by a marketplace, and weight above that cutoff results in a reduction of the price paid for the whole lamb. For example, if the cutoff was 125 pounds, and a lamb weighed 140 pounds, a penalty would be applied for the extra fifteen pounds. It is very much to the advantage of the producer to avoid such dockages by shipping earlier. Feed conversion is also generally poorer for large lambs. Remember that the weight that matters is the weight at the selling point, not the weight in your barn. Thus, lambs that are heavier than the cutoff by a few pounds can be safely shipped because they will lose the weight before weighing. Also, weights are generally average weights because a bunch of lambs are run onto the scale together, weighed, counted, and an average weight determined. You can mix in some lightweights to balance the heavies, so that the average weight stays below the cutoff. You don't want the disparity between the lightest and the heaviest to be too great, because that will reduce the uniformity of the lot, but a certain amount of variation is tolerable.

Just to confuse things, some packers will pay a premium for heavy lambs. Check your local situation.

I hope all this convinces you to buy a scale. If overweight dockage is costing you a dollar or three per lamb, it doesn't take very many lambs sent at acceptable weights to pay for the scale. If the idea of all the messing with the lambs and weighing them bores you, think again. Selling the lambs is the finale of your time and effort. Don't scrimp on your work now and reduce your profit for lack of a little time spent in the barn or pen. All those sleepless nights during lambing are wasted if you don't get top dollar for your products. Sure, you may get a good price for your wool and sell some breeding stock, but market lambs are your bread and butter. After you have weighed lambs for a few years you will find that you can estimate the weights pretty closely, so look at all the weighing as partly training to make you a better guesser.

All of this does not mean that you cannot ship a mixed lot of lambs, but have it made up so it can be conveniently divided into smaller lots for selling. You might ship some wethers at an average weight of 130 pounds, some ram lambs at 115 pounds, and some ewes and wethers at 110 pounds. Make it easy for the commission man to sort them into groups, and you'll get a better price. Be especially sure that they all grade Choice or Prime so he doesn't have to take out unfinished ones for sale as feeders.

With finished lambs, be sure the commission man knows that they will be ready to sell. Ideally, you should accompany some lambs to market to meet your representative and to get acquainted with what that market requires or expects. Ask a lot of questions, and you will learn a lot. Once the commission man knows you, and finds that you can be relied upon to deliver a good product, then you can be assured of getting the best price he can muster for you.

Once you have established your reputation with him, all you have to do is phone ahead to tell your representative what you are sending and to ask what price you can expect to get. You may also get advice as to whether to hold them a few days for a better market. A person who is selling and buying lambs every day can give you lots of help, so use him.

Some terminal markets are real auctions in which all dealing is in the open for anyone to observe. Others have all business conducted by what is called private treaty, in which the seller and buyer negotiate on a one-to-one basis until a mutually agreeable price is reached. The seller may also be negotiating with other buyers so that a private treaty sale can be almost as competitive as an auction. One possible conflict in the private treaty arrangement, or even in some auctions, is that the commission man may represent both seller and buyer. If this is the case at your marketplace, you can only hope that the person representing you is the type who can play both sides of the game without prejudice. Some of these arrangements seem to work satisfactorily, at least most of the time.

Grading Lambs

Grading lambs is not a complicated procedure, and it is something that every sheep raiser should take the time to learn. Even if you raise only a lamb or two a year for your own table, you should be competent to decide when it is ready for slaughter, or else you'll be getting it overfat and wasty or maybe have a bag of bones in your freezer.

There are two factors in grading a lamb for market. One of these is the condition of the pelt. An unshorn lamb, in the strict sense, means that it has never been shorn. However, in the trade, a lamb with wool longer than an inch and a half will usually be graded as unshorn, even if it actually was shorn once. Shorn lambs will have number-one pelts ($\frac{3}{4}$-inch to $1\frac{1}{4}$-inch wool), number-two pelts ($\frac{1}{2}$-inch to $\frac{3}{4}$-inch), or number-three pelts ($\frac{1}{4}$-inch to $\frac{1}{2}$-inch). The length of the wool equals the shortest fibers on the main part of the pelt, not including the wool of the legs or belly. Some people use a dime to measure with because a dime's diameter is just short of $\frac{3}{4}$ inch, so if the wool will cover a dime everywhere it is a number-one pelt.

The other factor in lamb grading is judging the flesh and fat cover over the ribs, loin, and hips. By far the best way to learn grading of finish, as this is called, is to be instructed by a person who is qualified to grade lambs. This could be a producer, a lamb buyer, a 4-H kid, or a county fair judge. If you can't find a person to instruct you, you can do a passable job with some instructions. First of all, you need an average sort of hand, not either bony or pudgy. Make a fist of your average hand. If the lower part of the ribs of a lamb feels like the back of your hand or is better covered, the lamb grades Prime. If it feels like the fingers pressed together on the front of your fist, it grades Choice. If it feels like your knuckles, it will not grade Choice, so feed the lamb more or ship it as a feeder.

After making a preliminary decision based on rib cover, you should feel the area over the loin—the area on either side of the spine between the dock and the last ribs. With a blackfaced breed or its crosses, if the lamb is to be graded Prime you should not be able to feel the backbone prominently, if at all, when you place your palm over the loin. With whitefaced breeds, the spine can be a little less covered and still make Prime. With lambs that grade Choice, the backbone can be felt, but should not be prominent or knobby. No lamb that grades Choice or Prime should be bony over the hips, nor should the spine feel very bumpy when you run your thumb the length of the backbone. If you want a further check, feel the top of a rear leg. It should be full and round in a Prime lamb. Grading is a difficult skill to acquire merely from reading about it, but give it a try, and then ask your commission man to check your grading out and give you suggestions.

Do be aware that weight is not a factor in grading. A lamb can grade Prime at 80 pounds, or one could fail to make Choice at 140 pounds. Some breeds finish earlier, at lighter weights, than others. Also, the feeding program influences at what weight a lamb will finish. Some small breeds such as Cheviots, Finns, and Southdowns, put on a high-energy ration, can finish out at 70 pounds, whereas a Rambouillet out on the range might not be finished at 140. It is a matter of adjusting the feeding schedule to the breed to achieve the desired finish at the desired weight.

Cooperative Marketing

If you can get together with a group of other sheep raisers, you may be able to influence the market a little bit by virtue of the larger numbers you have to bargain with. A packer's buyer is much more interested in handling a semitrailer load of 400–440 lambs than he is your pickup load of 12, for obvious reasons. The least complex sort of cooperative marketing is to assemble a load of graded lambs with some other raisers and negotiate a price with a buyer. The buyer can be an entrepreneur or a packer buyer, most commonly the latter. There are so many variations on this scheme that it would be useless to try to cover them all. The important advantage here is that yardage and commission costs are saved, and the packer will pay more for a coherent lot of lambs than for a number of small groups. When and where the lambs are to be weighed and who pays for the trucking are the two salient points to be considered in making a deal. Remember that lambs will lose a lot of weight during transportation and in holding pens, so be wary of any deal in which they are moved or held for any length of time prior to weighing. It doesn't really make any difference if the weighing is done at the packing plant after travel and an overnight stand, but the price agreed upon should be high to reflect the weight loss, or shrink, that would take place between the farm and the scale. If there is no agreement on a reasonable estimate of shrink, weigh the lambs at the point of origin and compare that with the weights at the packers. Then you have some facts in front of all the parties. For receiving payments from the current government support program, you will need weights at the point of slaughter, so be sure that your lambs can be identified. Check with your local Farm Service Agency office for details. To find its location try www.fsa.usda.gov/edso/.

Electronic Auctions

Another, somewhat more elaborate, way to market lambs in truckload quantities is to go through a teleauction or electronic auction. In this arrangement, a company sets itself up in the business of selling loads of lambs to packers. Lamb producers in a given area either grade their lambs themselves or have them graded and give the information to an area coordinator, who organizes the load on paper and telephones the description of the load (weights, sex, breeds, type of pelts, grades, and other information) to the teleauction company. The auction is set up by phone lines and a central computer is used to do the bookkeeping during the selling period. After a few minutes, bidding closes and the high bidder is notified. Bidders do not know who is bidding on the load; at least they are not meant to know. The packer that gets the bid then designates a loading date and time when one of his trucks will pick up the load at one or more loading points established by the producers. On the chosen day, generally within a few days of the auction day, the producers deliver their lambs, the lambs are weighed and loaded, and the truck hauls them away. Producers' checks are mailed out promptly by the teleauction company, which in turn collects from the packer.

The big advantage of the teleauction is reduced costs to the producer and, usually, sales at a higher price than at most of the terminal markets, depending on the area of the country. As an example, a teleauction in the central part of the country, Cornbelt Lamb Teleauction of Baraboo, Wisconsin, charges less than a dollar per head for its services. In our area of western Minnesota, the cost to a producer who grades his own lambs is less than a dollar and a half per head plus whatever it costs to get the lambs to a loading point, generally less than fifty miles distant. The total cost includes the teleauction fee and a small fee per lamb to pay a coordinator, the loading point and weighing facility, and to cover expenses incurred by the coordinator, mainly phone calls. In the ideal case, producers call the coordinator with descriptions of their graded lambs. When a load is put together on paper, an auction is held and the load sold. The electronic-auction organization notifies the coordinator of the price and the loading date and time, and the information is passed on by phone to the producers. Producers bring their lambs to the loading point, where the coordinator supervises weighing and loading and telephones final weights and numbers to the teleauction office.

Naturally there are glitches in real life. Trucks are late, producers are late, some producers get mad and back out (and are not welcomed back with open arms—ever), mistakes are made in the paperwork, lambs are delivered that do not grade Choice or better and have to be sent back, and so it goes. Sometimes the high bid is not sufficient, and some sales are made by private treaty, or the lambs can be held for another sale at a later date. These deviations from the ideal scheme are just part of real life and should be accepted as problems to be solved on an individual basis. When an electronic auction works right, and even when it doesn't, the return to the producer is higher than in most other ways of marketing.

Healthy Competition

Each marketing possibility must be judged on the basis of your local situation and alternatives weighed one against the other. The best possible condition is to have two or more alternate ways to market lambs to keep healthy competition alive. In our area there are two terminal markets within a 150-mile radius, one local auction that emphasizes a lamb market and is trying to build its livestock trade, and a teleauction in full operation. The producer can list his lambs with the teleauction. If that sale doesn't materialize for some reason, there is still the option of shipping to one of the terminal markets or hauling the lambs to the local livestock yard for sale every Wednesday. We feel very fortunate to have a competitive marketplace that wants our lambs.

OLDER SHEEP

As sheep age, at some point they have to be culled from the flock to make way for replacements. Retirement for a ewe or ram generally means a trip to the Mexican border or Florida where they are eagerly sought after as bargain protein by Hispanic families. The market for culls is much less variable in terms of price than the somewhat volatile lamb market, and there are only minor seasonal variations.

Older sheep should be shipped whenever it is determined that they are no longer productive members of the flock. That can mean anytime, but it is wise to consider availability of feed and the sheep's condition. A well-fleshed old sheep will bring a better price than a bag of bones. In the middle of winter when stored feed is being used it is probably wise to ship the sheep regardless of condition. The feed needed to add weight and finish will likely cost more than any possible increase in market value. In contrast, a ewe that comes out of weaning in thin condition may be doing so at a time when pasture is in surplus, so it is better economics to have her eat a while on cheap food and put on a little extra flesh.

BREEDING STOCK

In an ideal world of sheep raising, all lambs are raised to be sold at fancy prices to an eager crowd of buyers who can hardly wait to buy them as breeding stock. Very few shepherds manage actually to live out this dream in its entirety, but selling breeders is a good way to make a bit more money from sheep raising.

Many producers sell breeding stock without even realizing it, because some ewes that are sent to market, even some old culls, are bought by sharpies and resold as breeding animals. The real crooks buy lots of cheap ewes and sort them out into look-alike flocks and feed them up to make them have a healthy and pleasing appearance. Then they seek out buyers with a cover story such as that they are forced to get rid of their beloved flock because of financial woes, a divorce, ill health, or any one of a million phony excuses. You cannot avoid supplying these scoundrels, but you can at least avoid buying from them. Be sure to warn newcomers to sheep about the existence of such con artists. Nothing will discourage a new shepherd faster than a flock full of ewes with bad bags, infertility, chronic disease, and the like.

You can legitimately sell your older ewes as breeding stock instead of shipping them as culls. You'll get a better price, and have the satisfaction of not having to send an old friend to a dog food plant or a Mexican butcher shop. If a ewe still has a good bag and is fertile, she may be just what someone is looking for. Let it be known that you routinely sell your less productive ewes every year, and you may find customers beating a path to your door. You can even sweeten the deal by exposing the ewes to a top quality ram and selling them as bred ewes for a little more money. To keep your reputation and a clear conscience, you'd be wise to guarantee that they are bred and that they will freshen with a good bag. If they don't, refund the buyer's money and ship them as culls. For both you and the buyer, it is a good deal.

The next rung up the social ladder for purveyors of breeding stock is selling young commercial ewes. Commercial ewes are usually crossbreds of some sort that are used to crank out market lambs. They can be of any breed, though it is currently popular in many parts of the country to use whitefaced ewes with a blackfaced, purebred ram on them to make growthy market lambs. A whitefaced ewe will produce a bigger wool clip than most blackfaced crosses, and the wool is generally of a higher quality as well. Lambs from the black on white cross have the growthiness of the sire breed plus the vigor of heterosis, or crossbreeding vigor.

Many shepherds specialize in producing commercial ewes as replacements for sheep raisers who prefer to buy rather than raise their own. Reputation is everything in this business, as I guess it is in any business. Honesty in dealing, high quality of ewes, and getting known are the touchstones of this endeavor. Honesty is something you can control, and the quality of the ewes is a matter of breeding and good management, factors you can also control. Getting known is a bit more difficult. If you are new to sheep raising, nobody knows either you or your sheep. The first thing to do, once you have some ewes to sell that you are proud of, is to advertise. This means listing your name or that of your farm in magazines and newspapers that will reach sheep raisers, or, more importantly, sheep buyers.

Most sheep purchasers are cautious people, as they should be, and they will want to see your name in print for a while before they even consider buying sheep from you. List yourself in a breeder's directory in a magazine, use classified ads, and you may even want to put a block ad in magazines now and then. Keep doing this, consistently, for a few years, and one year you will be pleasantly surprised when the potential buyers start writing you and phoning you about getting some of your sheep. If there is an annual ewe sale in your area, consign some of your best ewes to it. If they are really good ewes, and you can make sure that they are, your reputation will benefit. For a ewe sale, be sure to have the sheep neatly sheared if they are blackfaced and trimmed attractively if they are wool breeds. Have the lot uniform in size, breeding, and appearance. You will undoubtedly get more for them than you would from a packer, so don't be disappointed when they nevertheless go for a little less than you might have dreamed. If you stick with it, in a few years' time your reputation will grow and you will get more for your ewes. Be patient.

A bit higher up the social ladder of breeders is the sale of purebred and registered sheep. These animals almost always command a higher price than their mongrel and unpedigreed cousins.

Purebreds are a known product, at least to some degree, and you can justly charge more for them. A purebreeder is expected to cull his registered flock severely and adhere to breed standards. A good purebreeder will ship a lot of his lambs to slaughter if they don't meet the standards of the breed and his conscience. Some purebreeders sell registered individuals of outstanding character for one price and unregistered purebreds at a lower price as commercial ewes or rams. Many purebreeders maintain both a registered flock and a purebred but unregistered flock.

Breed association politics is a lot more important in the purebred business than in other sheep enterprises. The successful seller of costly breeding stock has to show sheep at the right shows and win often, as well as buying quality stock to improve the flock. Advertising is very important in the purebred business, as is active membership in sheep associations of various sorts as well as the breed association. Publicity is vital and should be actively sought. If you enjoy sheep more than you do people, then you probably should steer clear of dealing in big-time registered sheep because a large part of it is public relations. Just having top quality sheep isn't enough.

There are a few shepherds who specialize in unusual types of sheep, such as the less common purebreds or sheep with colored wool. The rules of this game are really the same as with raisers of any commercial or purebred sheep except that the market for the product, breeding stock, is smaller and more widely scattered. On the other side of the coin, there is less competition too, so one factor must be balanced against the other in making a decision to enter this field. Before you make a big commitment to such an enterprise, think it out and work slowly unless you have money to spare. A specialized operation can evolve from a commercial sheep farm as your reputation gets a chance to grow.

We played all of these games simultaneously, as do many sheep raisers. We had a commercial flock that we were gradually converting to a specialized wool type. Our commercial flock gave us most of our market lambs. We selected outstanding crossbred ewes from our lamb crop to sell as commercial breeders. In addition, we used colored rams on part of the flock to produce some black and gray sheep, the best of which we sold as breeders. We also maintained a nucleus flock of registered Lincolns for our own use in a crossbreeding program and to produce registered stock for sale. This sounds more complicated than it was, because the only time we really had to treat the subflocks separately was during breeding. For the rest of the year the flock was fully integrated, black and white, blueblood and peasant, living together in perfect harmony.

WOOL

As with selling live lambs or sheep, there is more than one way to sell wool. The simplest way is to hire a shearer and then sell the wool to him. Most shearers will buy wool, either on their own or as an agent for a large wool-buying firm or cooperative. This method is handy, but will usually net you less cash than other methods. Because it is almost no bother, though, this is an attractive way to solve the wool problem.

You can learn to shear your sheep yourself and save the cost of hiring someone, but equipment is expensive and you may find that your health or physique is not up to the task. You will still have to find a buyer, such as a company representative or even a hide and fur dealer, although they generally pay less than other buyers.

Wool Pools
Shepherds also band together and market wool cooperatively by forming what is called a wool pool. Members of a pool establish a set of rules and try to get a good price by marketing in lots large enough to fill one or more railroad boxcars or semitrailers. Pool members generally commit a given number of fleeces to the pool when the wool may still be on the backs of the sheep,

and pay an earnest-money deposit of fifty cents a fleece or so to guarantee that the promised fleeces will be delivered at some future date. The pool either makes a group decision or appoints one or more people to sell the wool. Bids may be called for or negotiations may be made with a single buyer or broker. Once an offer is accepted, a loading point and date are established and pool members bring their wool to be loaded. The wool is generally weighed at the time of loading. As with other cooperative ventures, the price received is generally higher than that which could be expected by an individual producer marketing wool alone. Many pools will set a bottom price and will not sell the wool unless that lower limit is exceeded. If, after some agreed upon date, the pool wool is not sold, members can take their wool away from the pool commitment and get their earnest money refunded. Most pools, however, tend to keep the wool in the pool until a satisfactory price is obtained, so if you join a pool, be prepared to store the wool for some time if the market is below the expectations of pool members.

I should mention that there are pools and pools. I mean by this that some pools have been established so long that they are really just wool buying companies rather than co-ops. As in politics, some are more responsive to the little person than others. You'll have to check your own area to discover what your options are. Our area sheep producers' association formed a pool from scratch, and it was a success for many years, so you might want to consider starting a new pool if there is no existing one.

Wool for Handspinners

Direct marketing to handspinners or selling to spinning and weaving shops is a small but vital part of the wool growing business. The single factor for success in this field is quality, and quality mostly means cleanliness. Handspinners want to do their spinning in their homes, not in a barn, and they understandably don't want wool that is full of vegetable matter like burrs, seeds, and alfalfa leaf, or filthy with dung or mineral dirt. If you want to compete in the handspinners' wool market, you will have to keep your wool clean, clean, clean. That means raising sheep outdoors away from bedding in a barn. It also means feeding hay in such a way that a lot of it doesn't end up in the neck wool and shoulder wool (see Horsepasture Mfg in Supplies in appendix 6 for one design), and it means keeping pastures free of noxious weeds that contaminate the fleeces. Shearing has to be done in a clean place and done carefully to avoid cutting fibers in half to make so-called second cuts. The

The county fair: the word on the building leaves no doubt about what's here.

fleeces must be skirted; that is, the leg, belly, and face wool should be removed as well as any tangled or cotted portions. All of this is more work, but the rewards are high since the skirted wool sells for up to ten or more times as much as what you could expect from a wool buyer for ordinary fleeces.

Naturally you will have to develop your own market just as you will when trying to sell lambs to consumers. As in that situation, you will have to get the word out by advertising, and be altogether uncompromising with yourself in selling only a first-rate product. A good reputation will give you a cadre of satisfied customers who will buy from you again and again, and will also tell their spinner friends about you.

People often ask what type of wool is best for handspinning. The answer is clean wool. Length, fiber diameter, and color are matters of taste, and every spinner is different from every other one. One individual might prefer a fine, short wool to make a garment to be worn next to the skin, whereas another might want a coarse, even hairy, fleece to use for a rug or a wall hanging. There

A handsome lineup of Hampshire rumps.

is no such thing as a type of fleece preferred by handspinners, except that it be free of contaminants. Probably the best course is to specialize in a given type of wool, and when you have figured out how to keep it clean, seek a market for it. To reach the right people you will have to advertise in handspinning and weaving magazines. If you live near a college or university, ads on bulletin boards are often effective. Before you try to sell handspinners' wool you should become a handspinner yourself. If you are not a spinner, or your spouse or other close friend isn't, you won't understand a spinner's needs, and you will make mistakes that might prove fatal to your budding business. You should know your customers and how to reach them. Above all, don't let an unsatisfied customer go without a prompt refund or replacement. If a purchaser is not satisfied with a fleece, that is her or his judgment, not yours. Dirt, manure, and vegetable matter like alfalfa leaf may be a part of your life, but they are not a part of your customer's, so look at your fleece with that in mind. A customer rightly expects a clean fleece, by her or his own definition of clean.

Some 4-H lambs relax after winning prizes for proud owners.

A youngster competing in open-class judging answers a judge's question.

Handspinning wool in a county fair sheep building always arouses the interest of passersby.

APPENDIX 3

About Sheep Drugs

With the exception of some sprays and dips and the like that are under the supervision of the Environmental Protection Agency (EPA), the agency that approves drugs and various other chemicals for use with sheep is the Food and Drug Administration (FDA), which is part of the Public Health Service, which is, in turn, part of the Department of Health and Human Services. Notice that the FDA has nothing at all to do with the Department of Agriculture (USDA), or the Federal Trade Commission, or the Customs Service. It has no relation at all to the Animal and Plant Health Inspection Service (APHIS) that is part of USDA and is concerned with the importation of sheep insofar as control of disease is concerned. It is APHIS that has control of the scrapie eradication program, for example.

APPROVED DRUGS

An approved drug is one for which extensive tests have been made to demonstrate that the material (1) does what it purports to do (control a disease for example); (2) does what it does without harm to the animal being treated; and (3) when used with with an animal does not constitute a risk to humans, especially with food-producing animals such as sheep. The testing is done by the manufacturer of the drug or its agents and cooperating institutions such as universities and trade associations, all of whom are called sponsors. When the FDA approves a drug it means only that the sponsors have submitted data that the FDA considers adequate to demonstrate that the drug is safe and effective for the indications listed on the label, when used in accordance with label instructions.

If you as a shepherd use an approved material according to the label, then you are probably safe from prosecution if an illegal residue of the drug is found in one of your lamb carcasses. Also, if you use an approved drug according to the label and a bunch of your lambs roll over dead as a result, you can attempt to get recourse in the courts from the manufacturer, the sponsors, and FDA. (I am using words like "probably" because any decision on an individual case is up to a court of law.)

On the other hand, if you used an approved drug in an unapproved manner or used an unapproved drug in a food-producing animal and illegal residues were discovered in the meat or milk, then the product would be subject to condemnation for use as food, and you could possibly be subject to regulatory action by the FDA, such as seizure of the food, issuance of an injunction, or prosecution for causing the residues. In addition, if the material was unapproved or used in an unapproved manner, you would have no recourse against anyone if adverse effects occurred because of your use.

If you read between the lines, you will recognize that the use of an unapproved drug or an approved one in an unapproved way will likely not get you into any trouble if it is used with a nonfood animal, because nobody really cares.

Prescription and Nonprescription Drugs

Many common drugs, such as penicillin G, tylosin, tetracyclines of several types, and sulfas are available on a nonprescription or over-the-counter (OTC) basis. These are drugs the FDA considers safe for the producer to use without close supervision of a veterinarian. They are generally materials that leave no tissue residues or are considered to be relatively free from residue problems when used by producers if directions are followed. Drugs are labeled for OTC sale only if adequate instructions for a non-veterinarian can be printed on the label. Just because these drugs are readily available on the shelf of a drugstore or farm supply store doesn't mean that you can use them in any way you like. Once again, if residues are found, you are the one whom the FDA or some other agency will come looking for. You must be able to demonstrate that you used the drug according to label instructions

Feed additives are normally approved or not approved by the USDA rather than the FDA. For example, chlortetracycline (Aureomycin) is approved as an over-the-counter feed additive. However, for sheep, monensin (Rumensin) is not approved as a feed additive, and a veterinarian cannot legally prescribe it for that purpose. Some feed additives such as selenium are regulated by the FDA, because they are subject to the so-called Delaney clause, legislation which requires FDA approval for any substance that is suspected of causing cancer under some conditions.

Many drugs are available only on the prescription of a veterinarian. All prescription drugs must carry the following statement on the label: "CAU-TION: Federal law restricts this drug to use by or on the order of a licensed veterinarian." Most of these are drugs that have a high potential for causing illegal residues or injuring the animal if used improperly. If you use prescription drugs that have been obtained from a veterinarian, be sure that you understand how to use them correctly. It is the veterinarian's responsibility to instruct you to use them in a proper manner, but that doesn't relieve you of the responsibility of fulfilling your part of the bargain—following the vet's instructions.

If a veterinarian provides a prescription drug to a producer and makes a reasonable effort to explain the dosage, route of administration, withdrawal time, and other appropriate information, and if the drug is an approved one that is used in the manner stated on the label or package insert, then the vet is probably protected from a lawsuit if the drug doesn't produce the desired results.

In addition, for a veterinarian to prescribe a drug there must be a veterinary/client relationship. Basically that means that a veterinarian is not supposed to prescribe a drug for use by any unknown person who walks into the office or phones. The veterinarian must take responsibility for making judgments about the needs of the animal, and the client must agree to follow the instructions. The veterinarian must either examine the animal(s) or have sufficient personal knowledge of the keeping and care of the animal(s). The veterinarian must also be available for follow-up examination if needed, or arrange some way for follow-up to be done by a competent, qualified person.

It is important to give the full course of treatment for as many days as recommended by the label or a veterinarian. Even if symptoms vanish, continue the full treatment. If there is no recommendation, continue treatment for one to two days after the symptoms cease to be seen.

Off-Label Use

An exception to the rule on both OTC and prescription drugs is called "extra-label" or "off-label" use, which is using animal or human health products in ways not specified on the label—for example, at higher doses or in different species. Sheep are often treated by off-label drug use, because there are not enough sheep to justify the manufacturer paying for the long testing required for approval for use with sheep. Off-label drug use requires a veterinarian's prescription. Drugs that are not approved for any animal or human use cannot be used off-label.

Off-label drug use is only allowed if the animal's health is threatened or if the animal is suffering. The use is only for treatment or preventive medicine. That means that off-label use is not allowed for purposes of production or re-production. For example, extra-label drug use for purposes such as weight gain, feed efficiency, or milk production is not allowed.

As with prescription drugs, a valid veterinarian/client/patient relationship must be established. Also it must be determined that there is no approved drug labeled to treat the condition, or that treatment at the recommended dosage would not be effective. A record of any animal given extra-label treatment must be maintained by the producer for two years. The withdrawal time prior to marketing an animal or animal product for food purposes must be significantly extended, as determined by the veterinarian using scientific information. Extra-label drugs can be administered in drinking water, but not in feed. Finally, the drugs should be stored in an appropriate manner as indicated on the label or by the veterinarian. Note that some drugs are not to be kept refrigerated.

Unapproved Drugs

Drugs are approved for a specific type of use for a specific species, and any other use is unapproved, by definition. However, there is a large class of substances that are called Generally Recognized as Safe (GRAS) that includes common things like table salt, sugar, vinegar, and a lot of drugstore items such as sodium bicarbonate, calcium carbonate, ascorbic acid, common vitamins, many mineral supplements, and the like. A long list of GRAS materials that are commonly added to feed is published by the FDA. The GRAS list includes most inorganic materials that are added to sheep and lamb feeds. A notable exception is ammonium chloride, which is added to lamb feeds to prevent urinary calculi. Ammonium chloride, contrary to widespread belief of many feed-mill operators and veterinarians, is an approved additive for sheep feeds. Selenium and its compounds are not on the GRAS list either, but sodium selenite or sodium selenate is an approved feed additive for sheep. Approval in this case is by the FDA, because selenium is on a list of chemicals that are treated as drugs by legislation—the Delaney clause again.

Many other materials are approved for one or more species, but not for sheep. Because of this, sheep producers are constantly tempted to use unapproved drugs for treatment of sheep and lambs. A big problem is that neither the safety nor the effectiveness of a given drug for sheep may be known. Just because something is safe and effective for a cow doesn't mean that it is for a sheep.

A recent law, effective in late 2000, would allow pharmaceutical companies to more easily obtain label approval for sheep, by using test results from other species. The bill would, among other things, provide tax incentives to drug companies and extend periods of market exclusivity to lower the cost of the approval process. One hopes that this means more drugs will be available for OTC or prescription use with sheep, and off-label use of said drugs will not be needed.

There are various informal ways of testing drugs, of course. These can range from your getting advice from your neighbor who was a grade-school dropout and who hasn't washed a hypodermic syringe in thirty years, to a veterinarian's getting advice from other veterinarians who have made informal clinical trials of a drug. If you believe something that your neighbor tells you, you'll have to do it taking into account your opinion of your neighbor, and with the understanding that you are taking more or less of a chance. You may end up with a dead animal, or the FDA representative ringing your doorbell, and you'll have not a leg to stand on.

On the other hand, a veterinarian has access to published reports by other veterinarians and researchers in animal health. The studies he reads may not be as exacting as those demanded by the FDA for approval, but they represent controlled research by qualified persons. Not only that, a veterinarian has training and experience as a guide in using an unapproved drug or using an approved one at an unapproved dosage or other variation.

If you want to use injectable levamisole instead of a drench or to use a paste-type horse wormer, get your vet to prescribe it. You may find occasionally that your veterinarian hesitates to prescribe or use an unapproved drug. While a vet is within his legal rights to use an unapproved drug or an approved one in an unapproved manner, he is also held responsible for any illegal residues found in food produced by the animal treated. The veterinarian can also be held responsible by the animal's owner and may be subject to civil (malpractice) suits. In other words, the owner of the animal gets off the hook, but the hapless vet is stuck with blame from all quarters.

Disapproved Drugs

There are a few drugs that are forbidden by the FDA because they are known to have safety hazards. Any product containing these substances is subject to regulatory action when found in interstate commerce. (The FDA, a federal agency, has authority to regulate only when a product is sold across state lines.) Also, some drugs that were approved at one time have had approval withdrawn when new evidence of their harmful effects appeared later. The use of diethylstilbestrol (DES) implants is a good example.

The following drugs are prohibited in food-producing animals under any circumstances: Chloramphenicol, Clenbuterol, Diethylstilbestrol (DES), Dimetridazole, Fluroquinolones (except as specifically approved), Furazolidone (except topical), Ipronidazole, Glycopeptides, Nitrofurazone (except topical), Sulfonamides in lactating dairy animals, except those specifically approved.

Another point about questionable drugs is that the FDA has no way to require that it be allowed to review drugs before they are put on the market. Because of this loophole in the law, it is sometimes years before a given drug comes to the FDA's attention. Once they know about a drug and determine that it is not in the GRAS category or previously approved, then they can require the paperwork to show the drug's safety and effectiveness. It may be assumed that there are drugs on the market which the FDA would not approve of if they knew about them. For this reason you should be suspicious of a drug that you've never heard of, perhaps made by a small manufacturer. Someone may be selling "snake oil" for sheep that was made in a basement or garage. It may be useless or even harmful, but the manufacturer can get away with selling it until the FDA gets wind of the operation, at which time the business may just fold and vanish. A New Jersey veterinarian was taken to court by the FDA for selling antibiotic mixtures by mail to veal farmers in a number of states even though the veterinarian had never examined the animals and the drug mixtures were not approved. Even a vet can be in the present-day patent medicine business until he's caught.

For additional information check out www.fda.gov/cvm/. See also www.saanendoah.com/ for some excellent information about drugs and many other aspects of goat health that can be applied to many sheep situations.

WEIGHTS AND MEASURES

Weights and measure with medications are often given in metric terms. Here are some helpers.

Approximate household measure equivalents:

> 1 teaspoon (tsp) = 5 ml (cc)
> 1 Tablespoon (Tbsp)= 15 ml (ca. 3 teaspoons), or ½ oz
> 1 ounce (oz.) = 30 ml (ca. 2 Tablespoons)
> 1 cup = 16 Tablespoons = 240 ml
> 1 liter (l) = 1,000 ml (cc) or 1.06 quarts

Doses for many medications and supplements are written in kilograms (kg) animal weight.

Some metric equivalents:

> 1 kg (1,000 grams) equals 2.2 pounds
> 1 pound = 0.454 kg, or 454 grams
> 3 standard adult aspirin tablets weigh approx. 1 g
> 1 mg = 1,000 ug or mcg (ug & mcg = micrograms)
> 1,000 mg = 1 gram (g)
> 1 ml (milliliter) is equivalent to 1 cc (cubic centimeter)

APPENDIX 4

Sheep Economics

Any endeavor, whether a hobby or a full-scale business, involves expenses, and it may sometimes bring in revenue. Raising sheep can produce income greater than expenses if a lot of things are done with foresight and intelligence. If a profit motive is not part of why you want to raise sheep, stop reading here. If you are curious, read on.

To decide whether sheep raising is something you want to get into, you should look at the costs involved and the potential for sales of your products: lambs, wool, and sheep. To some extent, the potential for profit is a function of what other income a sheep raiser has, because what might be an unacceptable expense to one person could be a tax shelter to another.

TAXATION

I will assume that you readers are not raising sheep as a corporation, but as individuals. In that situation, your expenses and income from the sheep enterprise are mixed with your personal tax obligations.

Any profits from the sheep become part of your gross income for income tax purposes. Profits from sheep, or any other self-run business, are added to any salary or other taxable income that you may have. Losses can also be subtracted—up to a vague point. In addition, bona fide farmers have a number of income tax advantages, not the least of which is relaxation of requirements for paying estimated tax in advance quarterly payments.

In addition, there may be property tax obligations that relate to your raising sheep. In some states you may pay tax on the sheep! On the other hand, you may have vastly reduced property taxes if you are a farmer and live on that farm.

I'll zip through these two subjects quickly, and suggest that you seek professional help if you don't sing along easily.

Income tax
In the sheep (or any) business there are expenses and receipts. The difference between the two is profit, and the profit is the taxable part that should be looked at with care and imagination.

First the bad news—land you own is not an expense as far as the tax authorities are concerned until you sell it, at which point the income from the sale can be reduced by the amount the land cost you. If you pay rent to use land, that is a current expense, and if you receive rent, that is current income. Property taxes on the land are normally a business expense. Sales of lambs, sheep, wool, or any other product are counted as income.

Expenses include almost everything you pay money for to support your sheep farm. The list is endless. Obvious things such as feed, veterinary, fuel, insurance, seed, and hired help are ordinary expenses. Farm-related telephone and vehicle use are deductible too. Cost of buildings, fences, and tractors may have to be treated differently. However, any sheep-raising expense can be subtracted from your receipts in some way.

In addition, some sales of livestock can be taxed as capital gains instead of ordinary income, making the taxation rate lower. You may be able to obtain tax credits on some of your investments in buildings and equipment.

Consult with a tax advisor with some experience in income tax for farmers. This does not just mean a tax preparer who happens to live and work in a farming community; it means someone who understands all the minutia and tricks of farm taxes.

Profit Making

In order to use business deductions from a sheep-raising operation, you must be able to demonstrate to the Internal Revenue Service (IRS) that you are engaged in the activity with the intention of making a profit. The clearest proof is to make a profit every year. If you can't do that, then try to make a profit for two years out of five. If you still can't do that, then maybe you should get a bookkeeper who will be able to make the books show a profit two years out of five, even if you didn't really do it. That can often be arranged by adjusting in which tax year you buy and sell. This is known as creative accounting and can be totally legal.

If you still can't show a profit, then you must be able to demonstrate to the IRS that you have made a reasonable effort to make a profit, even though you failed to do so. If your sole source of income is farming, then it can be simply assumed that you are trying to make a profit. If you have income of some sort other than from farming, then you will have to prove that the sheep business is more than a hobby. You do this by being able to show that you have tried to be a good manager, that you are improving your knowledge, and are consulting experts and changing methods that have proven not to be profitable in the past, and similar activities. The fact that you are reading this book would be a piece of such evidence. Not only that, you can deduct the cost of the book as a business expense.

Enough of this—see that tax advisor with experience in income tax for farmers.

Fixed Costs

Some costs of raising sheep are called fixed costs. Why fixed? Because they are costs that don't change much from year to year or with the number of sheep one owns.

Important fixed costs are land, fences, buildings, and much equipment. Land is a one-time cost, and in general it doesn't deteriorate in value over time. Most other fixed costs change with time, because the item gets old and worn out. A fence is a fixed cost, but after some time period it will have to be replaced, so it declines in value. A convenient way to fit fixed costs into a budget is to estimate how much money you could earn from the fixed cost if you invested the money somewhere. Use current interest rates to do your figuring. As I write this, one can get about 6 percent interest on conservative investments. So, if you pasture your sheep on 40 acres of land that you paid $500 an acre for, then your annual cost is 40 × $500 × 0.06 = $1,200. Going one step further, if you have 200 adult ewes on the pasture, the annual cost is $1,200/200 = $6 per ewe.

Of course you should have a fence around the pasture, and some cross fencing within the 40 acres too. Your cost for that could be about $2,000, allowing some extra for maintenance and loss of interest over a life of twenty years. So that is $100 per year or $0.50 per ewe.

If you have a barn for lambing, that might cost you $5 a square foot for a 1,600-square-foot simple, basic barn or $8,000. Again over 20 years the cost per ewe will be $8,000/20/200 = $2 per ewe, ignoring the $0.25 in lost interest.

All the above come to $8.50 per ewe per year. One should allow for some other fixed costs, so round up by another $0.50 per ewe per year to give a total of $9 per ewe per year.

The ewe has a cost of course. Let's say you pay $100 per ewe, and then sell her for $20 after seven years. You lose $6 interest a year, so the annual cost per ewe is $6 + ($80/7) = about $18.

There is a ram cost, but a ram can breed up to 50 ewes a year with ease, so his cost will be less than $1 per ewe. Veterinary and drug costs are typically about $2.50 a year per ewe.

The sheep will eat from the pasture for half the year, but assuming six months of hay and some grain at critical times figure 600 pounds of hay at $0.03 per pound and 250 pounds of grain at $0.05 a pound for a total feed cost of $30.50 per ewe per year.

So we have:

Land, fencing, housing	$9.00
Feed	$30.50
Ewe cost	$18.00
Ram cost (per ewe)	$1.00
Health	$2.50
Other	$1.00
Total per ewe per year	$62.00

That ewe will produce about 8 pounds of wool a year. In the present wool market, the value of the wool may not pay for the shearing, so that could be a loss. If one raises sheep that produce premium, niche market wool, then the wool can be worth $4–$8 a pound after throwing away the dirty or spoiled parts, leaving 6 pounds or $24–$48 a fleece. However, shearing will cost $2–$6, and marketing will also be costly in both time and cash outlays for advertising or travel to direct markets. For this discussion I will just avoid the wool value entirely as it is subject to too many variables.

However, that ewe will also have lambs that can be marketed, and that is where the profit comes in. To look at that part of the spectrum, one must convert to cost per lamb. The cost per lamb depends on how many lambs a ewe births. Here are some possibilities.

number of lambs/ewe	cost per lamb at birth
1	$62.00
1.5	$41.33
2	$31.00
2.5	$24.80
3	$20.66

A lamb will eat some creep feed prior to weaning, perhaps averaging a pound a day over an eight-week period, for 56 pounds at $0.06 or $3.36. After weaning it should gain a pound on four pounds or less of grain. Assuming a light weaning weight of 50 pounds, to gain an additional 60 pounds would take 240 pounds of feed at $0.06 or $14.40 per lamb plus the preweaning $3.36 to give a total lamb feed cost of $17.76. You'd have to use local prices for feed to have this figure mean anything for your flock.

Using the above figures, which would be typical for the midwestern USA, we get the following. I have approximately corrected the feed costs for the heavier birth weights of the single, 1.5, and 2 lamb births. The correction is about $0.25 per pound of lamb.

lambs/ewe	feed to 110 pounds	cost per lamb
1	$17	$79.00
2	$18	$49.25
3	$19	$39.66

Let's assume a conservative $0.70 a pound for market lambs, and I will ignore marketing costs. Then the income per ewe and net profit would be as follows:

lambs/ewe	cost of all of a ewe's lambs	value of all of a ewe's lambs	net profit per ewe
1	$79.00	$77.00	(−$2.00)
2	$98.50	$154.00	$55.50
3	$118.98	$231.00	$112.02

I have not figured labor into any of the above calculations.

Why would anyone raise sheep that produced an average of one lamb per ewe? That's a good question. However, in a range operation the cost of land, feed, and housing is greatly reduced, and the total number of sheep is much larger than in a farm flock. The actual budget for one large ranch in New Mexico shows a total cost of $38.59 per ewe, including rental of pasture, feed, vet, interest, and everything else. So that operation would show a tidy profit with single lambs.

At the other end, a flock averaging three lambs per ewe will require a great deal more labor and skill from the shepherd, and would not be a good thing at all for inexperienced people. Also, with that many lambs the cost of labor begins to become very significant. A part-time farmer would probably have to hire help, increasing costs.

Calculations by Janet McNalley, a leading Minnesota producer and consultant, show that the most profitable range of lambing percentages is in the 180 to 225 percent range for average sheep and average shepherds. With prolific sheep who are also good milkers and mothers, and with skilled shepherding, higher lambing levels are very practical.

If one can sell breeding stock, maybe even a few expensive purebreds, and also find a good niche market for specialty wool, then the margins go up too.

MILK SHEEP

A small but growing niche market is sheep milk and cheese. There is a good book on the subject by Olivia Mills (see appendix 6), and there is a good booklet available from the Alfa Laval company called "Systems Solutions for Dairy Sheep." There is a big investment in housing and milking equipment, and planning is essential. As with milking cows, milking sheep is a seven-day-a-week job. Before planning or building any facilities, be sure to check with your state's milk inspectors to find what the requirements are. No matter what the regulations, you will need lots of water (both cold and hot), a sewage system, a bulk tank with cooling, and a place to wash equipment. You will also need facilities for storing milk in a large freezer to accumulate a large enough quantity to ship to a cheesemaker or other buyer.

The breed of choice is the East Friesian (or Friesland), and there are sheep with East Friesian bloodlines available in the United States. The problem is that East Friesian sheep cannot be imported from their homelands in Europe for health reasons, but there are sources in New Zealand and Australia for importable sheep, embryos, and semen. A recommended source for information is Dr. Martin Dally at the University of California, Davis. His email address is mrdally@ucdavis.edu. This is a developing field, so check current magazines for new information. Also check www.sheepdairying.com and olivia_mills @msn.com.

CAVEATS

So get out a pad and pencil, or fire up the spreadsheet software, and start experimenting with numbers before you buy the sheep, or even the land.

Don't forget, be sure that there is a market for your lambs in your area or region before you embark on raising lambs. Transporting lambs to a distant marketing point can wipe out profits with the flick of a pen on a hauling contract. Even if you have a convenient market, check out all the marketing costs and take them into account—which I did not in the material above. The same applies to wool, milk, or any other product from your sheep farm.

APPENDIX 5

Nutritional Requirements

S heep and lambs require water, a source of energy, a source of protein, and small amounts of vitamins and minerals. In most parts of the world, a pasture with a stream or pond will meet all of a sheep's nutritional needs, especially if a thoughtful shepherd has provided some salt with minerals in it. Requirements are very different at different times of the year, and a shepherd who pastures sheep year-round has to manage grazing in such a way as to prevent sheep obesity at some times and their near starvation at others. Requirements depend largely on the stage of the ewes' reproductive cycles. Where stored feed is used for part of the year, the amounts fed must be adjusted to suit the cyclic needs of ewes.

Water must always be in adequate supply or it will be a limiting factor in the whole nutritional picture. The same is true for minor and trace minerals and for vitamins, especially vitamins A and E that can be in short supply when inferior stored feed is used at some seasons. Recommendations for appropriate levels of vitamins and minerals are given in the tables that follow.

When most shepherds speak of nutrition they are referring to the energy and protein provided by feedstuffs such as hay, grass, and grain. The basic information about the gross nutritional needs of sheep, as well as the composition of common feeds, has been gathered into a book, *Nutrient Requirements of Sheep*, published by the National Research Council (NRC). The current sixth edition was published in 1985. Some tables from the book are reproduced here for your convenience (see tables 1 to 4 in this appendix), but I strongly recommend that you buy the whole booklet too, because it contains much more useful material than can be reproduced here. You can purchase or read the book on-line—see appendix 6. Please note that the tables are presented mostly in metric weights: kilograms and grams. In the sample calculations, I will convert from metric to English. The basic conversion factors are 1 kilogram (kg) = 2.2 pounds, 1 pound = 454 grams (g), 1 ounce = 28.4 g.

ENERGY

The energy requirements of sheep are that fraction of nutrient needs that is measured in calories, with table values given in millions of calories or Mcal. Unlike humans in developed countries, sheep generally receive insufficient energy rather than insufficient protein, in cases in which any dietary deficit is present. The energy content of feeds is given in terms of total digestible nutrients (TDN) or in energy units in Mcal.

The optimal levels of energy that a sheep requires depend on a number of factors, including air temperature, availability of shelter, age and sex, stage of the ewe's cycle, and general health. The NRC tables take some of these factors into account. The shepherd will have to use judgment and experience for other factors not tabulated.

A sheep that does not receive sufficient energy will lose body weight and fat, stop growing, produce less wool and milk, and be more susceptible to disease than an adequately nourished sheep. A pregnant ewe that does not get sufficient energy will give birth to premature and/or weak lambs, and the lambs will have a poor survival rate. A ram that does not get enough feed energy will not have the stamina or interest to get out and breed enough ewes to earn his keep, and the ewes will have low conception rates even if bred.

A ewe that receives too much energy will not be as fertile, may have difficult births because of oversized lambs and her own extra fat, will be more susceptible to vaginal prolapse during late gestation, will produce less milk when lactating, will readily go off feed from indigestion, and will be much more susceptible to a variety of diseases than her less portly sister. An overfed ram will simply not be fit enough to get out and do his job without getting pooped.

The commonest concentrated energy feed is grain, although pasture, hay, and root crops can provide lots of calories too. A situation to avoid is feeding a material, straw, for example, that is too low in TDN relative to its volume. A sheep in some cases cannot consume enough material to meet its minimum needs. Silage can pose a similar problem if high water content keeps a sheep from eating enough to meet peak needs. Just plain palatability can also be a problem with spoiled, moldy, or weedy hay.

PROTEIN

As with energy needs, protein requirements are easily met by quality pasture or good hay. Except for really poor feeds, the protein requirements are almost automatically satisfied if energy needs are met. This statement should not be taken literally to the extent that protein needs are ignored because some feeds are practically all energy and almost no protein. Protein requirements of lambs are vastly higher than those of older ewes, so consider age and stage in the reproductive cycle.

Feeds such as soybean meal, sunflower seed meal, rapeseed (canola) meal, and similar products are very high in protein, and are usually used to blend high protein lamb feeds. Good quality legume hay will average 12 to 16 percent protein, far more than sufficient for an adult ewe at any stage in her cycle. Grains, as mentioned, are primarily energy feeds and have only around 8 to 10 percent protein for common feed types. Urea, a nonprotein chemical that contains nitrogen, is sometimes used as a nitrogen source to increase the effective protein content of mixed feeds, because the rumen flora can convert small proportions of it into protein building blocks. It is very tricky to use, and poor mixing can make a portion of the feed deadly to ewes eating it. In any case, no more than one-third of the total "protein" can be in the form of urea, and it can be fed only to functional ruminants. Young lambs, less than a month old or so, cannot utilize urea at all, and will be sickened or killed by it.

The emphasis on protein in human diets leads some sheep producers to be overly concerned with it. Sheep convert protein in feed into protein in their own bodies insofar as it is needed. Any excess protein has to be converted into metabolizable energy for the sheep to be able to use it. This is a losing proposition because the sheep must use up energy in order to convert the extra protein to energy, a sort of robbing Peter to pay Paul situation. The economy-minded shepherd should be sure to feed enough protein, but avoid overfeeding it, because it is utilized very inefficiently as a source of energy. You will find when you calculate some rations that it is sometimes difficult to devise a mix that is low enough in protein.

MINERALS

Calcium and phosphorus are the main minerals required by sheep in terms of quantity. The calcium and phosphorus should be in a ratio of about 2:1 (Ca:P) and should total about 0.5 percent of the whole diet. Legumes are generally very high in calcium relative to phosphorus, with cereal grains being the reverse. Grass hays are intermediate. Sheep can tolerate high Ca:P ratios, up to six or seven to one, but the reverse is not true.

Sodium is usually adequate in ordinary feed, and in any case is provided amply in salt given free choice. Some self-styled experts say there is no need for supplemental salt for sheep, but try telling that to your flock. Whether they need it or not is a moot point because they surely do like it. As I mentioned in chapter 7, adequate sodium from salt is essential for lactation. Trace minerals are easily provided by free choice TM salt as long as copper toxicity is not a problem. Never mix a copper-containing mineral salt with feed, because the sheep may consume a toxic amount.

CALCULATING RATIONS

There are many computer programs (software) for calculation of livestock rations. Depending on the sophistication of the software, the final ration may be a minimum-cost mixture of a large number of components or just a balanced blend of two or three materials.

For most purposes, simple ration calculations can be done with a pencil and paper and a pocket calculator. I will give a couple of examples to get you started.

Example 1—Calculating a Ration from Available Hay and Grain for a 70 kg (154 lb.) Ewe in Late Gestation Carrying Twins

Available feeds are a grass hay and shelled corn that have analyses as listed below, with protein, digestible energy (DE), Ca, P on a dry basis.

Check the following table, keeping in mind that a 70 kg ewe in late gestation requires 5.4 Mcal of DE, 214 g protein, 7.6 g Ca, and 4.5 g P.

Feed Analyses

	% Dry Matter	% Protein	DE, Mcal/kg	% Ca	% P
Grass hay	90	8.1	2.47	0.34	0.21
Shelled corn	85	10.0	4.05	0.02	0.32

The ewe will be able to eat about 2 percent of her body weight in hay with ease, or 1.4 kg (roughly 3 pounds). The 1.4 kg of hay will provide 1.4 × 2.47 = 3.46 Mcal on a dry basis. Since the hay is 90 percent dry matter, the energy of 1.4 kg becomes 3.46 × 0.90 = 3.11 Mcal.

The ewe requires 5.4 Mcal, so the corn will have to supply 5.40−3.11 = 2.29 Mcal. The corn contains 4.05 Mcal/kg, so the ewe needs 2.29 / 4.05 = 0.56 kg. Since the corn contains only 85 percent dry matter, we should divide by 0.85 to get 0.66 kg (about 1.5 pounds) of corn as fed.

Her energy requirements then will be met by 3 pounds of the hay and 1.5 pounds of the corn.

The protein in the ration will be:

Hay	1.4 kg	× 0.081	× 1,000	× 0.90 =	102 g
Corn	0.66 kg	× 0.100	× 1,000	× 0.85 =	56 g

Total protein = 158 g

The ewe needs 214 g of protein, so she is 214−158 = 56 g short. This deficit can be provided out of a bag by feeding some soybean meal along with the corn. Soybean meal contains about 46 percent protein as fed, so 56 / 0.46 = 122 g or a little more than four ounces. Perhaps a simpler route might be to feed a higher protein hay since even a 12 percent protein hay would make up the protein need.

Calcium provided by the basic ration will be:

Grass hay	1.4 kg	× 0.90	× 0.34% Ca × 1,000	= 4.28 g Ca
Shelled corn	0.66 kg	× 0.85	× 0.02% Ca × 1,000	= 0.11 g Ca

Total Ca = 4.39 g

A similar calculation for phosphorus gives total P = 4.4 g.

The slight deficit in phosphorus would be easily provided for by giving the ewe free-choice access to a 1:1 salt and dicalcium phosphate mixture. Dicalcium phosphate contains almost 19 percent P, so the mixture would contain about 9 percent P. Typical daily consumption of salt mixes is 10−15 g, and even 10 g would provide 1.2 g Ca and 0.9 g P. The dicalcium phosphate could be mixed with the grain at the rate of 5 g per ewe if desired.

The calcium deficit could be provided by feeding alfalfa hay or by adding ground feed-grade limestone (33 percent Ca) to the grain at 10 g per ewe. The ewe is getting 0.66 kg of corn so the 10 g of limestone would be 1.5 percent of the total grain mixture, or 30 pounds per ton of corn.

A ewe carrying triplets should be fed more, but not so much as to make her put on fat (perhaps 10 percent additional TDN).

Example 2—Calculating a Lamb Grain Ration with 14 Percent Protein Using Corn and Soybean Meal with, Respectively, 9 Percent and 46 Percent Protein as Fed

To solve this problem, simply calculate the difference between the protein content of each ingredient and the desired total protein content of the mix, without regard to algebraic sign.

	% Protein	Desired % Protein	Difference
Corn	9	14	5
Soybean meal	46	14	32

The value of the difference is the proportion of the *other* ingredient. Thus the desired proportions of corn and soybean meal to give a 14 percent mixture are 32 parts corn and 5 parts soybean meal. In percentage terms that would be corn 86 percent, soybean meal 14 percent; or a ton of feed could be made of 1,720 pounds of corn and 280 pounds of soybean meal. Rounding down or up to the nearest hundred pounds of each component will give:

Corn	Soybean Meal	%Protein
1,700	300	14.6
1,720	280	14.0
1,800	200	12.7

The calcium and phosphorus requirements of young and growing lambs are most important and should be considered. A 40 kg (88 lb.) ewe lamb requires 5.9 g Ca and 2.6 g P.

Each kg of the corn-soybean meal mixture will contain:

	%	Ca	P
		$0.86 \times 0.02\%$	$0.86 \times 0.32\%$
Corn	86	$\times 1,000 = 0.17g$	$\times 1,000 = 2.75g$
Soybean meal	14	$0.14 \times 0.36\%$	$0.14 \times 0.75\%$
		$\times 1,000 = 0.50g$	$\times 1,000 = 1.05g$
		Total Ca 0.67g	Total P 3.80g

A kilogram of the ration (2.2 lb.) will easily provide the required phosphorus but is markedly deficient in calcium. In addition, the Ca:P ratio should be around 2:1 as a preventive measure against urinary calculi. Addition of 1 percent calcium carbonate (limestone) will add 3.4 g/kg of calcium to give a total of 4.07 g/kg. Addition of 2 percent limestone will bring the ratio to almost 2:1, as well as providing 6.8 g/kg additional calcium.

The calcium can also be provided by an alfalfa hay with 1.3 percent calcium. To provide the same 6.8 g as the ground limestone would do, the lamb will have to eat 6.8 / 0.013 = 523 g (over a pound) of hay for each kilogram of the grain-meal mixture. This is not at all unreasonable, except that the limestone is still cheap insurance against a low Ca:P ratio for the lamb that finds the corn and soymeal mixture more to its liking and eats little or no hay. Getting extra calcium from the hay will do no harm, so providing the appropriate ratio in the concentrated feed is a reasonable thing to do.

Table 1 Daily Nutrient Requirements of Sheep

Body Weight (kg)	(lb.)	Weight Change/Day (g)	(lb.)	Dry Matter per Animal (kg)	(lb.)	(% body weight)	Energy TDN (kg)	(lb.)	DE (Mcal)	ME (Mcal)	Crude protein (g)	(lb.)	Ca (g)	P (g)	Vitamin A Activity (IU)	Vitamin E Activity (IU)
Ewes[c]																
Maintenance																
50	110	10	0.02	1.0	2.2	2.0	0.55	1.2	2.4	2.0	95	0.21	2.0	1.8	2,350	15
60	132	10	0.02	1.1	2.4	1.8	0.61	1.3	2.7	2.2	104	0.23	2.3	2.1	2,820	16
70	154	10	0.02	1.2	2.6	1.7	0.66	1.5	2.9	2.4	113	0.25	2.5	2.4	3,290	18
80	176	10	0.02	1.3	2.9	1.6	0.72	1.6	3.2	2.6	122	0.27	2.7	2.8	3,760	20
90	198	10	0.02	1.4	3.1	1.5	0.78	1.7	3.4	2.8	131	0.29	2.9	3.1	4,230	21
Flushing—2 weeks prebreeding and first 3 weeks of breeding																
50	110	100	0.22	1.6	3.5	3.2	0.94	2.1	4.1	3.4	150	0.33	5.3	2.6	2,350	24
60	132	100	0.22	1.7	3.7	2.8	1.00	2.2	4.4	3.6	157	0.34	5.5	2.9	2,820	26
70	154	100	0.22	1.8	4.0	2.6	1.06	2.3	4.7	3.8	164	0.36	5.7	3.2	3,290	27
80	176	100	0.22	1.9	4.2	2.4	1.12	2.5	4.9	4.0	171	0.38	5.9	3.6	3,760	28
90	198	100	0.22	2.0	4.4	2.2	1.18	2.6	5.1	4.2	177	0.39	6.1	3.9	4,230	30
Nonlactating—First 15 weeks gestation																
50	110	30	0.07	1.2	2.6	2.4	0.67	1.5	3.0	2.4	112	0.25	2.9	2.1	2,350	18
60	132	30	0.07	1.3	2.9	2.2	0.72	1.6	3.2	2.6	121	0.27	3.2	2.5	2,820	20
70	154	30	0.07	1.4	3.1	10	0.77	1.7	3.4	2.8	130	0.29	3.5	2.9	3,290	21

Table 1 Continued

80	176	30	0.07	1.5	3.3	1.9	0.82	1.8	3.6	3.0	139	0.31	3.8	3.3	3,760	22
90	198	30	0.07	1.6	3.5	1.8	0.87	1.9	3.8	3.2	148	0.33	4.1	3.6	4,230	24

Last 4 weeks gestation (130–150% lambing rate expected) or last 4–6 weeks lactation suckling singles[d]

50	110	180 (45)	0.40 (0.10)	1.6	3.5	3.2	0.94	2.1	4.1	3.4	175	0.38	5.9	4.8	4,250	24
60	132	180 (45)	0.40 (0.10)	1.7	3.7	2.8	1.00	2.2	4.4	3.6	184	0.40	6.0	5.2	5,100	26
70	154	180 (45)	0.40 (0.10)	1.8	4.0	2.6	1.06	2.3	4.7	3.8	193	0.42	6.2	5.6	5,950	27
80	176	180 (45)	0.40 (0.10)	1.9	4.2	2.4	1.12	2.4	4.9	4.0	202	0.44	6.3	6.1	6,800	28
90	198	180 (45)	0.40 (0.10)	2.0	4.4	2.2	1.18	2.5	5.1	4.2	212	0.47	6.4	6.5	7,650	30

Last 4 weeks gestation (180–225% lambing rate expected)

50	110	225	0.50	1.7	3.7	3.4	1.10	2.4	4.8	4.0	196	0.43	6.2	3.4	4,250	26
60	132	225	0.50	1.8	4.0	3.0	1.17	2.6	5.1	4.2	205	0.45	6.9	4.0	5,100	27
70	154	225	0.50	1.9	4.2	2.7	1.24	2.8	5.4	4.4	214	0.47	7.6	4.5	5,950	28
80	176	225	0.50	2.0	4.4	2.5	1.30	2.9	5.7	4.7	223	0.49	8.3	5.1	6,800	30
90	198	225	0.50	2.1	4.6	2.3	1.37	3.0	6.0	5.0	232	0.51	8.9	5.7	7,650	32

First 6–8 weeks lactation suckling singles or last 4–6 weeks lactation suckling twins[d]

50	110	−25 (90)	−0.06 (0.20)	2.1	4.6	4.2	1.36	3.0	6.0	4.9	304	0.67	8.9	6.1	4,250	32
60	132	−25 (90)	−0.06 (0.20)	2.3	5.1	3.8	1.50	3.3	6.6	5.4	319	0.70	9.1	6.6	5,100	34
70	154	−25 (90)	−0.06 (0.20)	2.5	5.5	3.6	1.63	3.6	7.2	5.9	334	0.73	9.3	7.0	5,950	38
80	176	−25 (90)	−0.06 (0.20)	2.6	5.7	3.2	1.69	3.7	7.4	6.1	344	0.76	9.5	7.4	6,800	39
90	198	−25 (90)	−0.06 (0.20)	2.7	5.9	3.0	1.75	3.8	7.6	6.3	353	0.78	9.6	7.8	7,650	40

Table 1 Daily Nutrient Requirements of Sheep (continued)

Body Weight (kg)	Body Weight (lb.)	Weight Change/Day (g)	Weight Change/Day (lb.)	Dry Matter per Animal (kg)	Dry Matter per Animal (lb.)	Dry Matter per Animal (% body weight)	Energy TDN (kg)	Energy TDN (lb)	DE (Mcal)	ME (Mcal)	Crude protein (g)	Crude protein (lb)	Ca (g)	P (g)	Vitamin A Activity (IU)	Vitamin E Activity (IU)
First 6–8 weeks lactation suckling twins																
50	110	−60	−0.13	2.4	5.3	4.8	1.56	3.4	6.9	5.6	389	0.86	10.5	7.3	5,000	36
60	132	−60	−0.13	2.6	5.7	4.3	1.69	3.7	7.4	6.1	405	0.89	10.7	7.7	6,000	39
70	154	−60	−0.13	2.8	6.2	4.0	1.82	4.0	8.0	6.6	420	0.92	11.0	8.1	7,000	42
80	176	−60	−0.13	3.0	6.6	3.8	1.95	4.3	8.6	7.0	435	0.96	11.2	8.6	8,000	45
90	198	−60	−0.13	3.2	7.0	3.6	2.08	4.6	9.2	7.5	450	0.99	11.4	9.0	9,000	48
Ewe lambs																
Nonlactating—First 15 weeks gestation																
40	88	160	0.35	1.4	3.1	3.5	0.83	1.8	3.6	3.0	156	0.34	5.5	3.0	1,880	21
50	110	135	0.30	1.5	3.3	3.0	0.88	1.9	3.9	3.2	159	0.35	5.2	3.1	2,350	22
60	132	135	0.30	1.6	3.5	2.7	0.94	2.0	4.1	3.4	161	0.35	5.5	3.4	2,820	24
70	154	125	0.28	1.7	3.7	2.4	1.00	2.2	4.4	3.6	164	0.36	5.5	3.7	3,290	26
Last 4 weeks gestation (100–120% lambing rate expected)																
40	88	180	0.40	1.5	3.3	3.8	0.94	2.1	4.1	3.4	187	0.41	6.4	3.1	3,400	22
50	110	160	0.35	1.6	3.5	3.2	1.00	2.2	4.4	3.6	189	0.42	6.3	3.4	4,250	24
60	132	160	0.35	1.7	3.7	2.8	1.07	2.4	4.7	3.9	192	0.42	6.6	3.8	5,100	26
70	154	150	0.33	1.8	4.0	2.6	1.14	2.5	5.0	4.1	194	0.43	6.8	4.2	5,950	27

Table 1 Continued

Last 4 weeks gestation (130–175% lambing rate expected)

40	88	225	0.50	1.5	3.3	3.8	0.99	2.2	4.4	3.6	202	0.44	7.4	3.5	3,400	22
50	110	225	0.50	1.6	3.5	3.2	1.06	2.3	4.7	3.8	204	0.45	7.8	3.9	4,250	24
60	132	225	0.50	1.7	3.7	2.8	1.12	2.5	4.9	4.0	207	0.46	8.1	4.3	5,100	26
70	154	215	0.47	1.8	4.0	2.6	1.14	2.5	5.0	4.1	210	0.46	8.2	4.7	5,950	27

First 6–8 weeks lactation suckling singles (wean by 8 weeks)

40	88	−50	−0.11	1.7	3.7	4.2	1.12	2.5	4.9	4.0	257	0.56	6.0	4.3	3,400	26
50	110	−50	−0.11	2.1	4.6	4.2	1.39	3.1	6.1	5.0	282	0.62	6.5	4.7	4,250	32
60	132	−50	−0.11	2.3	5.1	3.8	1.52	3.4	6.7	5.5	295	0.65	6.8	5.1	5,100	34
70	154	−50	−0.11	2.5	5.5	3.6	1.65	3.6	7.3	6.0	301	0.68	7.1	5.6	5,450	38

First 6–8 weeks lactation suckling twins (wean by 8 weeks)

40	88	−100	−0.22	2.1	4.6	5.2	1.45	3.2	6.4	5.2	306	0.67	8.4	5.6	4,000	32
50	110	−100	−0.22	2.3	5.1	4.6	1.59	3.5	7.0	5.7	321	0.71	8.7	6.0	5,000	34
60	132	−100	−0.22	2.5	5.5	4.2	1.72	3.8	7.6	6.2	336	0.74	9.0	6.4	6,000	38
70	154	−100	−0.22	2.7	6.0	3.9	1.85	4.1	8.1	6.6	351	0.77	9.3	6.9	7,000	40

Replacement ewe lambs[e]

30	66	227	0.50	1.2	2.6	4.0	0.78	1.7	3.4	2.8	185	0.41	6.4	2.6	1,410	18
40	88	182	0.40	1.4	3.1	3.5	0.91	2.0	4.0	3.3	176	0.39	5.9	2.6	1,880	21
50	110	120	0.26	1.5	3.3	3.0	0.88	1.9	3.9	3.2	136	0.30	4.8	2.4	2,350	22
60	132	100	0.22	1.5	3.3	2.5	0.88	1.9	3.9	3.2	134	0.30	4.5	2.5	2,820	22
70	154	100	0.22	1.5	3.3	2.1	0.88	1.9	3.9	3.2	132	0.29	4.6	2.8	3,290	22

Table 1 Daily Nutrient Requirements of Sheep (continued)

Body Weight (kg)	(lb.)	Weight Change/Day (g)	(lb.)	Dry Matter per Animal (kg)	(lb.)	(% body weight)	Energy TDN (kg)	(lb.)	DE (Mcal)	ME (Mcal)	Crude protein (g)	(lb.)	Ca (g)	P (g)	Vitamin A Activity (IU)	Vitamin E Activity (IU)
Replacement ram lambs[e]																
40	88	330	0.73	1.8	4.0	4.5	1.1	2.5	5.0	4.1	243	0.54	7.8	3.7	1,880	24
60	132	320	0.70	2.4	5.3	4.0	1.5	3.4	6.7	5.5	263	0.58	8.4	4.2	2,820	26
80	176	290	0.64	2.8	6.2	3.5	1.8	3.9	7.8	6.4	268	0.59	8.5	4.6	3,760	28
100	220	250	0.55	3.0	6.6	3.0	1.9	4.2	8.4	6.9	264	0.58	8.2	4.8	4,700	30
Lambs finishing—4 to 7 months old[f]																
30	66	295	0.65	1.3	2.9	4.3	0.94	2.1	4.1	3.4	191	0.42	6.6	3.2	1,410	20
40	88	275	0.60	1.6	3.5	4.0	1.22	2.7	5.4	4.4	185	0.41	6.6	3.3	1,880	24
50	110	205	0.45	1.6	3.5	3.2	1.23	2.7	5.4	4.4	160	0.35	5.6	3.0	2,350	24
Early weaned lambs—Moderate growth potential[f]																
10	22	200	0.44	0.5	1.1	5.0	0.40	0.9	1.8	1.4	127	0.38	4.0	1.9	470	10
20	44	250	0.55	1.0	2.2	5.0	0.80	1.8	3.5	2.9	167	0.37	5.4	2.5	940	20
30	66	300	0.66	1.3	2.9	4.3	1.00	2.2	4.4	3.6	191	0.42	6.7	3.2	1,410	20
40	88	345	0.76	1.5	3.3	3.8	1.16	2.6	5.1	4.2	202	0.44	7.7	3.9	1,880	22
50	110	300	0.66	1.5	3.3	3.0	1.16	2.6	5.1	4.2	181	0.40	7.0	3.8	2,350	22

Table 1 Continued

Early weaned lambs—Rapid growth potential[f]

10	22	250	0.55	0.6	1.3	6.0	0.48	1.1	2.1	1.7	157	0.35	4.9	2.2	470	12
20	44	300	0.66	1.2	2.6	6.0	0.92	2.0	4.0	3.3	205	0.45	6.5	2.9	940	24
30	66	325	0.72	1.4	3.1	4.7	1.10	2.4	4.8	4.0	216	0.48	7.2	3.4	1,410	21
40	88	400	0.88	1.5	3.3	3.8	1.14	2.5	5.0	4.1	234	0.51	8.6	4.3	1,880	22
50	110	425	0.94	1.7	3.7	3.4	1.29	2.8	5.7	4.7	240	0.53	9.4	4.8	2,350	25
60	132	350	0.77	1.7	3.7	2.8	1.29	2.8	5.7	4.7	240	0.53	8.2	4.5	2,820	25

[a]To convert dry matter to an as-fed basis, divide dry matter values by the percentage of dry matter in the particular feed.

[b]One kilogram TDN (total digestible nutrients) = 4.4 mcal DE (digestible energy); ME (metabolizable energy) = 82% of DE. Because of rounding errors, values in Table 1 and Table 2 may differ.

[c]Values are applicable for ewes in moderate condition. Fat ewes should be fed according to the next lower weight category and thin ewes at the next higher weight category. Once desired or moderate weight condition is attained, use that weight category through all production stages.

[d]Values in parentheses are for ewes suckling lambs the last 4–6 weeks of lactation.

[e]Lambs intended for breeding; thus, maximum weight gains and finish are of secondary importance.

[f]Maximum weight gains expected.

Source: Data from National Research Council, *Nutrient Requirements of Sheep*, 6th ed. (Washington, D.C.: National Academy of Sciences, 1985)

Table 2 Macromineral Requirements of Sheep (Percentage of Total Diet Dry Matter)

Nutrient	Requirement
Sodium	0.09–0.18
Calcium	0.20–0.82
Phosphorus	0.16–0.38
Magnesium	0.12–0.18
Potassium	0.50–0.80
Sulfur	0.14–0.26

Source: Data from National Research Council, *Nutrient Requirements of Sheep,* 6th ed. (Washington, D.C.: National Academy of Sciences, 1985)

Table 3 Micromineral Requirements of Sheep and Maximum Tolerable Levels (ppm, mg/kg of Total Diet Dry Matter)

Nutrient	Requirement	Maximum Tolerable Level
Iodine	0.10–0.80	50
Iron	30–50	500
Copper	7–11[a]	25[b]
Molybdenum	0.5	10[b]
Cobalt	0.1–0.2	10
Manganese	20–40	1,000
Zinc	20–33	750
Selenium	0.1–0.2	2
Fluorine	———	60–150

[a]Requirement when dietary molybdenum is less than 1 mg/kg. For higher Molybdenum levels more may be required, but that is rare.

[b]Lower levels may be toxic under some circumstances. See source, pp. 16–18 for details.

Source: Data from National Research Council, *Nutrient Requirements of Sheep,* 6th ed. (Washington, D.C.: National Academy of Sciences, 1985)

Table 4 Composition of Some Common Sheep Feeds

sc = sun cured	Dry Matter (%)	DE Sheep (Mcal/kg)	ME Sheep (Mcal/kg)	TDN Sheep (%)	Protein (?)	Digest. Protein (%)	Calcium (%)	Phosphorous (?)
Alfalfa hay, sc, late vegetative	90	2.29	1.88	52	17.9	14.3	1.38	0.26
Alfalfa hay, sc, early bloom	90	2.22	1.82	51	16.2	12.7	1.27	0.20
Alfalfa hay, sc, full bloom	90	2.10	1.72	47	13.5	9.4	1.13	0.20
Alfalfa hay, sc, mature	91	2.17	1.78	49	11.7	7.7	1.03	0.17
Barley, grain	88	3.35	2.74	76	11.9	9.8	0.04	0.34
Barley, straw	91	1.93	1.58	43	4.0	0.7	0.27	0.07
Beet, mangel, fresh root	11	0.39	0.32	9	1.3	0.9	0.02	0.02
Beet, sugar, dehydrated	91	2.96	2.43	67	8.8	4.5	0.63	0.09
Bluegrass hay, Poa compressa, sc	92	2.44	2.00	55	9.6	4.1	0.28	0.25
Bromegrass hay, sc	91	2.20	1.80	49	8.8	4.8	0.31	0.17
Calcium phosphate (dical)	97	-	-	-	-	-	21.3	18.7
Canarygrass, reed, hay	91	1.97	1.61	45	9.4	5.8	0.35	0.23
Clover, red, hay, sc, early bloom	86	2.44	2.00	55	16.0	11.4	1.24	0.30
Corn, grain (54 lb/bushel)	88	3.39	2.78	77	8.9	5.7	0.02	0.31
Corn, silage, with ears	33	1.03	0.85	23	2.7	1.2	0.08	0.07
Corn, stalks, no ears/husks, stover	85	2.21	1.82	50	5.6	2.5	0.49	0.08
Cottonseed meal, mech. extracted	93	2.58	2.12	59	37.9	27.3	0.20	0.90
Limestone, ground	98	-	-	-	-	-	33.32	0.02
Meadow plants, intermtn, hay, sc	95	2.43	1.99	55	8.3	5.0	0.58	0.17

Table 4 Composition of Some Common Sheep Feeds (continued)

sc = sun cured	Dry Matter (%)	DE Sheep (Mcal/kg)	ME Sheep (Mcal/kg)	TDN Sheep (%)	Protein (?)	Digest. Protein (%)	Calcium (%)	Phosphorous (?)
Oats, hay, sc	91	2.13	1.75	49	8.5	5.2	0.22	0.20
Oats, straw	92	1.91	1.57	44	4.1	0.3	0.22	0.06
Orchardgrass hay, sc	91	2.33	1.91	53	10.2	6.5	0.35	0.32
Prairie plants, midwest, hay	95	2.26	1.86	51	6.7	2.7	0.34	0.12
Rapeseed meal, mech. extracted	92	3.04	2.49	69	35.6	30.1	0.66	1.04
Rye, distillers grain, dried	92	2.59	2.13	59	21.6	12.9	0.05	0.48
Sorghum, grain	90	3.48	2.85	79	11.1	8.2	0.03	0.29
Sorghum, silage	30	0.74	0.61	17	2.2	0.6	0.10	0.06
Soybean meal, mech. extracted	90	3.37	2.77	77	42.9	36.8	0.26	0.61
Sunflower seed meal, solv. extr.	90	1.79	1.46	41	23.3	18.9	0.21	0.93
Sweetclover hay, sc	87	2.04	1.67	46	13.7	10.0	1.11	0.22
Timothy hay, midbloom	89	2.35	1.93	53	8.1	4.9	0.43	0.20
Trefoil, birdsfoot, hay, sc	92	2.36	1.93	53	15.0	10.3	1.57	0.25
Turnip, root, fresh	9	0.35	0.29	8	1.1	0.8	0.05	0.02
Wheat, bran	89	2.78	2.28	63	15.2	11.8	0.11	1.22
Wheat, grain	89	3.41	2.80	78	14.2	11.4	0.04	0.37
Wheat, hay, sc	88	2.01	1.65	45	7.4	4.0	0.13	0.17
Wheat, straw	89	1.60	1.32	36	3.2	-3.1	0.16	0.04
Wheatgrass, crested, hay, sc	93	2.17	1.78	49	11.5	7.4	0.31	0.20

Source: Data from National Research Council, *Nutrient Requirements of Sheep*, 6th ed. (Washington, D.C.: National Academy of Sciences, 1985)

APPENDIX 6

Sources

SHEEP INDUSTRY ORGANIZATIONS

American Sheep Industry Association (ASI)
6911 S. Yosemite St.
Englewood, CO 80112
303-771-3500, fax 303-771-8200
www.sheepusa.org

National Lamb Feeders Association
   (NLFA)
1270 Chemeketa St. NE
Salem, OR 97301

National Sheep Association
950 S. Cherry St., Ste. 508

Denver, CO 80246
303-758-3513, fax 303-758-0190
www.nationalsheep.org
celliot@agri-associations.org

OPP Concerned Sheep Breeders Society
Holly Neaton, DVM, Secretary-
   Treasurer
11549 Highway 25 SW
Waterton, MN 55388
952-955-2596
hollyneat@juno.com
www.interrain.com/opp

SHEEP BREED ASSOCIATIONS

Barbados Blackbelly
Blackbelly Barbados Sheep Association
   International
2050 Griffith Ave.
Terrell, TX 75160
972-551-7090
info@blackbellysheep.org
www.blackbellysheep.org

North American Barbados Blackbelly
   Sheep Registry
P.O. Box 237
McKean, PA 16426
barbadoshp@aol.com
members.aol.com/barbadoshp

Black Sheep
American Black Sheep Registry
Carol Bliss, Secretary
4714 Glade Rd.
Loveland, CO 80538
970-667-0208
bmerino@concentric.net
www.concentric.net/~bmerino/
   absr.html
All colored sheep accepted.

Black Welsh Mountain
American Black Welsh Mountain Sheep
   Association
P.O. Box 534
Paonia, CO 81428-0534

oogiem@desertweyr.com
www.blackwelsh.org—for general association info
www.blackwelsh.net—for info on the black welsh mail list

Blueface Leicester
Blueface Leicester Union of North America
Kelly Ward, Secretary
760 West V.W. Ave.
Schoolcraft, MI 49087
616-679-5497
wardfarm@net-link.net
www.bflsheep.com

Hexham-X Registry
for Bluefaced Leicester crosses
P.O. Box 604
Norco, CA 92860
909-734-7307
khornerstn@aol.com
www.kpmcornerstone.com/hexhamx/index.htm

Booroola Sheep
(no formal association, just a contact)
Janet McNally
Tamarack Lamb and Wool
RR 2 Box 63
Hinckley, MN 55037
320-384-7262
tamarack@pinenet.com

Border Leicester
American Border Leicester Association
(since 1973)
Diana Waibel, Secretary
P. O. Box 947
Canby, OR 97013

503-266-7156
dgw@gggnet.com
www.interrain.com/abla

American Border Leicesters, Ltd.
P. O. Box 133
McLemoresville, TN 38235
901-742-4536
borderleicesters@att.net

North American Border Leicester Association
Deb Deakin, Registrar
P.O. Box 500
Cuba, IL 61427-9625
309-785-5058 (day), fax 309-785-5050

California Red
California Red Sheep Registry
Janice Altomare, Secretary
1850 Reilly Rd.
Merced, CA 95340
209-725-0340
highseas@cell2000.net
www.cell2000.net/ca_redsheep

California Variegated Mutant
see: American Romeldale/CVM Association under Romeldale

Cheviot
American Cheviot Sheep Society
Ruth Anne Bowles, Secretary
RR 1, Box 100
Clarks Hill, IN 47930
765-523-2767
Mark Bernard mbernard@smig.net
members.aol.com/culhamef/bcheviots/cheviot.htm

Clun Forest
North American Clun Forest
 Association
Bets Reedy, Secretary
21727 Randall Drive
Houston, MN 55943-9748
507-864-7585
bramble@acegroup.cc
www.clunforestsheep.org

Columbia
Columbia Sheep Breeders Association
 of America
Richard Gerber, Secretary
P.O. Box 272
Upper Sandusky, OH 43351
740-482-2608
csbagerber@udata.com
www.columbiasheep.org

Coopworth
Coopworth Sheep Society of North
 America
Marcia Adams, Secretary
25101 Chris Ln NE
Kingston, WA 98346-9303
360-297-4485
rainfarm@tscnet.com

Cormo
American Cormo Sheep Association
Charlotte Carlat, Secretary-Treasurer
RR 59
Broadus, MT 59317
406-427-5449
carlat@mcn.net

Corriedale
American Corriedale Association
Marcia Craig, Secretary

P.O. Box 391
Clay City, IL 62824-0391
618-676-1046

Cotswold
American Cotswold Record Association
Vicki Rigel, Secretary
18 Elm St, P.O. Box 59
Plympton, MA 02367-0059
781-585-2026
ACRAsheep@aol.com

The Black Cotswold Society
Brenda Leppo, Registrar
P.O. Box 1158
North Plains, OR 97133
503-647-5710
Contact: Pat LaBreque
 pearl@totalnetnh.net

Cotswold Breeders Association
Julie Mangnall, Secretary
21092 478th Ave.
Bushnell, SD 57276-6504
605-693-4354
mangnall@itctel.com
Anthony Kaminsky, Registrar
P.O. Box 441
Manchester, MD 21102

Debouillet
Debouillet Sheep Breeders Association
P.O. Box 67
Picacho, NM 88343
505-653-4084 or 505-622-0055
Contact: Mary Skeen

Delaine-Merino
American and Delaine Merino Record
 Association

Elaine Clouser, Secretary
1026 County Rd. 1175, Rte. 3
Ashland, OH 44805
419-281-5786
info@admra.org
www.admra.org

Texas Delaine Association
Mrs. G. A. Glimp, Secretary
Rt 1, Box 26
Burnet, TX 78611
512-756-4257

Blacktop Delaine Merino Association
John Mater, President
10641 E. State Rd.
Nashville, MI 49073-0000
517-852-9247

Black-Top Delaine Merino Sheep
    Breeders' Association
1775 Damman Rd.
Fowlerville, MI 48836
517-223-9728

Black Top and National Delaine
    Merino Sheep Association
RD 4, Box 228-F
McDonald, PA 15057
412-745-1075

Dorper
American Dorper Sheep Breeders'
    Society
Ronda Sparks, Registrar
1120 Wilkes Road
Columbia, MO 65201
573-442-4103, fax 573-874-8843
RondaS@sockets.net
www.dorperamerica.org

Dorpcroix
Dorpcroix Sheep Farm & Registry
Bill Hoag
6585 CR 4105
Hermleigh, TX 79526
915-863-2775
dorpcroix@aol.com
hometown.aol.com/dorpcroix/
    myhomepage/business.html

Dorset
Continental Dorset Club
Debra Hopkins, Secretary
P.O. Box 506
North Scituate, RI 02857-0506
401-647-4676, fax 401-647-4679
cdcdorset@dellnet.com

Finnsheep
Finnsheep Breeders Association
Elizabeth Luke, Secretary
HC 65, Box 495
DeRuyter, NY 13052
315-852-3344
stillmeadowfinns@hotmail.com
www.finnsheep.org

Gulf Coast Native
Gulf Coast Sheep Breeders Association
Pat Piehota, Secretary
Rt 2 Box 43A
Snyder, OK 73566
580-569-2631
gcsba@juno.com

Hampshire
American Hampshire Sheep Association
Karey Claghorn, Secretary
1557 173rd Ave. #8
Milo, IA 50166

641-942-6402, fax 641-942-6502
info@hampshires.com
www.countrylovin.com/ahsa/index.html

Icelandic
Icelandic Sheep Breeders of North
   America
P.O. Box 427
Gorham, ME 94038-0427
207-793-4640
secretary@isbona.com
www.isbona.com

Jacob
Jacob Sheep Breeder's Association
Lane Harris, Secretary
P.O. Box 10427
Bozeman, MT 59719
information@jsba.org
spottedsheep@yahoo.com
www.jsba.org

Jacob Sheep Conservancy & Breed
   Registry
1165 E. Lucas Rd.
Lucas, TX 75002-7455
972-727-0900
jhorak@earthlink.net
www.jacobsheepconservancy.org

Karakul
American Karakul Sheep Registry
Rey Perera, Registrar
11500 Hwy 5
Boonville, MO 65233
660-838-6340, fax 660-838-6322
aksr@mid-mo.net
www.karakulsheep.com

Katahdin
Katahdin Hair Sheep International
Teresa Mauer, Operations
P.O. Box 778
Fayetteville, AR 72702-0778
501-444-8441 (day), fax 501-444-8441
khsint@earthlink.net
www.khsi.org
edmartsol@mev.net

Lincoln
National Lincoln Sheep Breeders
   Association
Roger Watkins, Secretary-Treasurer
1152 CTH H
Mt. Horeb, WI 53572
608-437-5086
watkins@mhtc.net
Registration and transfers done by
Karey Claghorn of the Hampshire
   Association.
1557 173rd Ave. #8
Milo, IA 50166
641-942-6402, fax 641-942-6502
worldsheep@aol.com
www.lincolnsheep.org

Montadale
Montadale Sheep Breeders Association
Mildred Brown, Secretary
P.O. Box 603
Plainfield, IN 46168
317-839-6198
pjbus@aol.com
www.montadales.com

Natural Colored
Natural Colored Wool Growers
   Association
Barbara Kloese, Registrar

429 West US 30
Valparaiso, IN 46385
219-759-9665, fax 219-759-9665
kloese@gte.net
www.ncwga.org

Navajo Churro
Navajo-Churro Sheep Association
Connie Taylor, Registrar
P.O. Box 94
Ojo Caliente, NM 87549
505-737-0488
churro@newmex.com
www.navajo-churrosheep.com

North Country Cheviot
American North Country Cheviot
  Sheep Association
Edward Racel, Secretary
8708 S CR 500 W
Reelsville, IN 46171-0000
765-672-8205
yuccafl@ccrtc.com

Oxford
American Oxford Sheep Association
Mary Blome, Secretary
1960 E. 2100 North Rd.
Stonington, IL 62567
217-325-3515

Panama
(No formal association, just a contact)
Troy Ott
University of Idaho, AVS Dept
Moscow, ID 83843-0000
208-855-6075 (day), fax 208-855-6420
jmiller@uidaho.edu

Perendale
Perendale Breeders Association
Norlaine Schultz, Secretary
18811 New Hampshire Ave.
Ashton, MD 20861
301-774-0484

Polypay
American Polypay Sheep Association
Debbe Anderson, Secretary
609 S. Central Ave., Ste 9
Sidney, MT 59270
406-482-7768
polypay@lyrea.com

Rambouillet
American Rambouillet Sheep Breeders
  Association
1557 173rd Avenue
Milo, IA 50166
641-972-6402, fax 641-942-6502

Romanov
North American Romanov Sheep
  Association
Don Kirts, Secretary
P.O. Box 1126
Pataskala, OH 43062-1126
740-927-3098
narsa@netwalk.com
www.netwalk.com/~narsa

Romeldale
American Romeldale/CVM Association,
  Inc.
31808 79th Ave. Ct. E
Eatonville, WA 98328
253-846-5863
Contact: Sharon Hayden
mark@blue-moon-farm.com

Romney
American Romney Breeders Association
Jean Kamenicky, Secretary-Treasurer
744 Riverbanks Road
Grants Pass, OR 97527
541-476-6428
jkkorral@internetcds.com
www.americanromney.org

Scottish Blackface
Scottish Blackface Sheep Breeders
    Association
Richard Harward, Secretary
1699 H H Hwy
Willow Springs, MO 65793
417-962-5466

Shetland
North American Shetland
    Sheepbreeders Association
Julie Guilette, Secretary
265 Truway Road
Luxemburg, WI 54217-9559
920-837-2167
bramblewool@itol.com
www.shetland-sheep.org

Shropshire
American Shropshire Registry
    Association
Dale E. Blackburn, Secretary
24905 Streit Rd., P.O. Box 635
Harvard, IL 60033-0635
815-943-2034, fax 815-943-2034

Southdown
American Southdown Breeders'
    Association
Gary Jennings, Secretary
HCR 13, Box 220

Fredonia, TX 76842
915-429-6226, fax 915-429-6225
southdown@ctesc.net

St. Croix
St. Croix Hair Sheep Association
Jo Van Hoy, Secretary
Box 845
Rufus, OR 97050
509-773-5988 or 435-797-2181
blackpackranch@pocketmail.com
Cole Evans, Pres.-Registrar
Utah State Univ., UMC 9400
Logan, UT 84322-4815
801-797-2181 (day)
cevans@mendel.usu.edu

Suffolk
Ontario Suffolk Sire Reference
    Association
Joe Stephenson, Secretary
8093 Fuller Rd.
Forest, ON
Canada N0M 2N0
519-243-1991
jrstephenson@odyssey.on.ca
ossra.tripod.com

United Suffolk Sheep Association
Missouri Office
Ronda Sparks, Office Manager
3316 Ponderosa St.
Columbia, MO 65201
314-442-4103, fax 314-443-3632
rondas@sockets.net
Utah Office
Annette Benson, Secretary
17 W. Main, P.O. Box 256
Newton, UT 84327-0256
435-563-6105, fax 435-563-9356

suffolksheep@pcu.net
www.u-s-s-a.org

Targhee
U.S. Targhee Sheep Association
Christine Ashmead, Secretary
P.O. Box 202
Fernwood, ID 83830
208-245-3869
ashmead@smgazette.com
www.ttc-cmc.net/~schuldt

Teeswater
American Teeswater Association
Barbara Kloese, Registrar
429 West U.S. 30
Valparaiso, IN 46385
219-759-9665 (voice & fax)
kloese@gte.net
Contacts: Barbara Burrows
ewesincolor@wyoming.com
www.angelfire.com/wy/teeswater
Myrtle Dow
myrtledow@aol.com

Texel
Texel Sheep Breeders Society
Bonnie Davis, Secretary
24001 N 1900 E Road
Odell, IL 60460
815-998-2359, fax 815-998-2113
jbdavis@fcg.net

Tunis
National Tunis Sheep Registry
Judy Harris, Registry Clerk
819 Lyons St.
Ludlow, MA 01056
413-589-9653
pbrashea@icx.net
www.ezroane.com/ntsri/

Wensleydale
North American Wensleydale Sheep
  Association
Sherry Carlson, Secretary
4589 Fruitland Rd.
Loma Rica, CA 95901
530-743-5262
info@wensleydalesheep.org
www.wensleydalesheep.org

Wiltshire Horn
see: Am Livestock Breeds
  Conservancy
P.O. Box 477
Pittsboro, NC 27312
919-542-5704 or 919-545-0022
albc@albc-usa.org
www.albc-usa.org
contact: Pat Lenzo
patlenzo@surfsouth.com
www.psmag.com/HC/Wiltshires

You can check www.rbparker.com/sheeporganizations.html for updates of the above list. Please also notify ron@rbparker.com of any changes.

Stockmaster (UK) www.stockmaster.co.uk/site/welcome.asp lists British breeds and how to contact breed societies. Also links to the Rare Breeds Survival Trust (RBST).

Books, Video

An Introduction to Keeping Sheep, 2nd ed., 1996
Jane Upton and Dennis Soden.
Diamond Farm Book Pubns, ISBN 0-85236-332-X

A Practical Guide to Sheep Disease Management, 1985
Norman Gates
News-Review Publishing, Moscow, ID 52760-9999
ASIN 9-99960-022-0

Beginning Shepherds Manual, 1997
Barbara Smith
Iowa State University Press, ISBN 0-81382-799-X

Lambing Time Management
(set of three videotapes) by Don Baily, DVM and Woody Lane, PhD
Garden Valley Productions
240 Crystal Springs Lane
Roseburg, OR 97470
541-440-1926
wlane@rosenet.net

Managing Your Ewe, 1997
Laura Lawson
LDF Publications, ISBN 0-96339-231-X
Laura has two other books:
Lamb Problems, 1996, ISBN 0-96339-230-1
Showing Sheep, 1994, ISBN 0-96339-232-8
Order from Laura at 11114 Lawson Lane
Culpepper, VA 22701 ph. 703-825-0339

Mountain Sheep: A Study in Behavior and Evolution, 1971
Valerius Geist
Chicago University Press ISBN: 0-22628-572-3

Nutrient Requirements of Sheep, 6th ed, 1985
National Research Council
National Academy Press, ISBN 0-309-03596-1

read online free or buy at:
www.nap.edu/catalog/614.html

Practical Sheep Dairying: The Care and Milking of the Dairy Ewe, 1982
Olivia Mills
Thorsons or HarperCollins (paper), ISBN 0-72250-731-3

Sheep Housing and Equipment Handbook/MWPS-3, 1994
Harvey Himing, Tim Faller, Karl Hoppe
Midwest Plan Service, ISBN 0-89373-090-4

Sheep Raisers Manual, 1984
William K. Kruesi
Williamson Publishing, ISBN 0-91358-910-1

SID Sheep Production Handbook
American Sheep Industry Association
6911 South Yosemite
Englewood, CO 80112-1414
303-771-3500
View samples at:
www.sheepusa.org/resource/handbook/handbook.htm

Small-Scale Sheep Farming,1997
Jeremy Hunt
Faber and Faber, ISBN 0-57117-893-6
UK oriented.

Storey's Guide to Raising Sheep, 2000
Paula Simmons and Carol Ekarus
Storey, ISBN 1-58017-262-8

The Merck Veterinary Manual, 8th ed, 1998
Merck and Co., ISBN 0-91191-029-8

The Veterinary Guide for Sheep Farmers, 1990
David C. Henderson
Farming Press, ISBN 0-85236-189-0
Outstanding book.

Western Canadian Sheep Production Manual
(which includes the Nutrition Guide for BC Sheep Producers by Dr.
Steve Mason)
Available for about $10 from Alberta Sheep and Wool Commission:
 www.therockies.com/aswc

If you have trouble locating books from the United Kingdom try:
Diamond Farm Book Publishers
P.O. Box 537, Alexandria Bay, New York 13607
USA
or
Diamond Farm Book Publishers
RR 3 Brighton, ON
Canada K0K 1H0
800-481-1353 (Mon.–Fri. 8–5 EST)
613-475-1771
info@diamondfarm.com
www.diamondfarm.com/genstore.htm

Landsman's Bookshop, Ltd
Buckemhill, Bromeyard
Herefordshire HR7 4PH UK
01885-483429
sheep books at: www.landsmans.co.uk/f-files/a26.htm

MAGAZINES

Sheep Canada
Box 4, Site 8, RR#1
Olds, AB
Canada T4H 1P2
Cathy Gallivan, Ph.D.
403-224-3962, fax 403-224-3339
888-241-5124
gallivan@sheepcanada.com
www.sheepcanada.com

The Black Sheep Newsletter
25455 NW Dixie Mtn Rd

Scappoose, OR 97056
Peggy Lundquist
503-621-3063
bsnewsltr@aol.com
hometown.aol.com/jkbsnweb/

The Shepherd Magazine
5696 Johnston Rd.
New Washington, OH 44854-9736
Guy and Pat Flora
419-947-9289 or 419-492-2364
fax 419-947-1302
shepmag@bright.net

sheep! Magazine
P.O. Box 10
Lake Mills, WI 53551
Dave and Doris Thompson
920-648-8285, fax 920-648-3770
www.sheepmagazine.com/

The Shepherd's Journal
Box 99
Mossleigh, AB
Canada T0L 1P0
403-534-2185, fax: 403-534-2144
Garry & Doreen Schneider
shepherd@telusplanet.net
www.shepherdsjournal.com

The Working Border Collie
14933 Kirkwood Road
Sidney, OH 45365
937-492-2215, fax 937-492-2211
WBC@wesnet.com
www.working-border-collie.com/

Sheep Connection
2145 Megee Lane

Nicholasville, KY 40356
Dan and Susan Perkins
859-858-4622
suffolks@yahoo.com
www.kyagr.com/sheepconnection.htm
Somewhat regional (KY, TN, GA), but general information too.

Sheep & Goat Research Journal
P.O. Box 51267
Bowling Green, KY 42102-5567
Glenn Slack
phone/fax: 270-782-8370
sheep2goat@aol.com
Technical journal.

Supplies, Equipment

Canadian Co-Operative Wool Growers
Box 130, 142 Franktown Road
Carleton Place, ON
Canada K7C 3P3
613-257-2714, fax 613-257-8896
ccwghq@wool.ca
or
918 1st Avenue South
Lethbridge, AB
Canada T1J 0A9
800-567-3693, fax 403-380-6982
ccwgab@telusplanet.net
www.wool.ca/catalog.htm

Ash & Charlene Clements
342 Frontier Road
R D 6 Te Awamutu
New Zealand
Supplier of NZ eartags & NZ lamb teats
clements@wave.co.nz
www.clements.co.nz/

D-S Livestock Equipment
18059 National Pike
Frostburg, MD 21532
800-949-9997, fax 301-689-9727

Horsepasture Mfg
P.O. Box 25
Spencer, VA 24165
540-957-2558
Homespun@kimbanet.com
horsepasture.homestead.com
Hay feeders to keep wool clean.

Jeffers
P.O. Box 948
West Plains, MO 65775
800-JEFFERS

KV Vet
P.O. Box 245
3190 N. Road
David City, NE 68632
800-423-8211

Mid-States Livestock Supplies
9449 Basil Western Road, NW
Canal Winchester, OH 43110
800-835-9665
www.midstateswoolgrowers.com/asp/supplies.asp
or
125 E. 10th Avenue
South Hutchinson, KS 67505
316-663-5147 or 800-835-9665

NASCO
800-558-9596
www.nascofa.com/prod/Home
info@nasco.com
or

NASCO farm supply—2 locations
NASCO-Fort Atkinson-Eastern division
901 Janesville Ave.
Fort Atkinson, WI 53538-0901
920-563-2446, fax 920-536-8296
or
NASCO Modesto-Western division
4825 Stoddard Rd.
Modesto, CA 95356-9318
209-545-1600, fax 209-545-1669

Omaha Vaccine
3030 L Street
Omaha, NE
800-367-4444, fax 800-242-9447
catalog@omahavaccine.com
www.omahavaccine.com/

PBS Livestock Health Supply
P.O. Box 9101
Canton, OH 44711-9101
800-321-0235

Pipestone Vet Supply
1300 So Hwy 75, P.O. Box 188
Pipestone, MN 56164
Order only: 800-658-2523
(Catalog has helpful hints in it as well as a free phone-the-vet number on Fridays.)
Phone the vet 507-825-5687 Fridays 1:00 P.M. to 4:00 P.M. CST.

Premier Sheep Supplies, Ltd.
2031 300th St.
Washington, IA 52353
800-282-6631, fax 800-346-7992
info@premier1supplies.com
www.premier1supplies.com
(Premier's fencing catalog is almost a textbook on building electric fences.)

Sydell, Inc.
46935 SD Hwy 50

Burbank, SD 57010
888-848-4177
605-624-4538, fax 605-624-3233
sydell@sydell.com
www.sydell.com/

Townsend's Sales
4141 South 25 West
Trafalgar, IN 46181
317-736-4047

Valley Vet Supply
1118 Pony Express Hwy.
Marysville, KS 66508
Phone orders or questions: 800-360-4838
fax orders: 800-446-5597
service@Valleyvet.com
www.valleyvet.com/

Wiggins & Associates Inc.
1155 SW Towle Ave.
Gresham, OR 97080-9626
503-667-0716, fax 503-667-4701
800-600-0716
jpwiggins@att.net
www.wigginsinc.com
Has Meador's electric docker.

Wooltique
P.O. Box 537, 1111 Elm Grove St.
Elm Grove, WI 53122
orders: 800-657-0746
info: 414-784-3980

ARTIFICIAL BREEDING

Dr. Martin Dally
University of California, Davis
Davis, CA 95616
mrdally@ucdavis.edu

Elite Genetics
605 Rossville Road
Waukon, IA 52172
319-568-4551, fax 319-568-6370
lab 319-568-4847
email: eltgenetic@aol.com
www.elitegenetics.com

TANNING PELTS

Bucks County Fur Products
Box 204
220 ½ N. Ambler Street
Quakertown, PA 18951
215-536-6614

Moosehead Tanners
P.O. Box 1242
Moosehead Industrial Park
Greenville, ME 04441
207-695-0272

New Method Fur Dressing
131 Beacon St.
South San Francisco, CA 94080
Ph: 650-583-9881
Ask for Sandra.

Stern Tanning
P.O. Box 55, 334 Broadway
Sheboygan Falls, WI 53085
920-467-8615

SHEEP MANAGEMENT COMPUTER SOFTWARE

Ewe Byte
P.O. Box 375
Fergus, ON
Canada N1M 3E2
519-787-0593, fax 519-787-2675
ewebyte@entex.net
www.ovcnet.uoguelph.ca/associations/ewebyte/ewebyte.htm
Considered the best by many users.

Farmstock
Farmworks
P.O. Box 250, 6795 S. State Road 1
Hamilton, IN 46742-0250
800-225-2848
scott@farmworks.com
Not sheep specific.

Flockmaster
ABM
417-862-3353
Some negative comments.

Oviration
Software for feedstuff calculation.
www.softagro.com/oviration.html

Report Generation Services
2237 Glenwood Rd.
Vestal, NY 13850
Phone 607-785-2322
scubic@earthlink.net
home.earthlink.net/~scubic/
Has a sheep module and medical module, very simple level according to the supplier.

Shepherd Software
Mountain View Software
RR 1, Box 9, Site 9
Didsbury, AB

Canada T0M 0W0
403-335-9477
info@mountainviewsoftware.com
www.mountainviewsoftware.com

For enrollment in a national program see also:
National Sheep Improvement Program (NSIP)
6911 South Yosemite Street, Suite 200
Englewood, CO 80112-1414
303-771-5717, fax 303-771-8200
www.nsip.org
info@nsip.org

SOME USEFUL WEB SITES

Cornell University has www.sheep.cornell.edu/sheep/index.html and www.vet.cornell.edu/consultant/consult.asp dealing with husbandry and veterinary care. They also have a poisonous plant site at www.ansci.cornell.edu/plants

Purdue University has vet.purdue.edu/depts/addl/toxic/cover1.htm for more on poisonous plants.

The Maryland Small Ruminant Page, www.sheepandgoat.com/ has many useful links.

Check www.adds.org or phone Mike Opperman at 608-848-0955 for information about the National Sheep Database CD. The search system is rather crude and slow, but there is lots of information, including the SID Handbook.

From Australia comes a very useful veterinary site: www.sheepvet.com

The USDA has the following sites about cost-sharing programs: www.fsa.usda.gov/pas/ and www.ams.usda.gov/lsg/

For animal drug information from FDA: informatics3.vetmed.vt.edu/

For FDA CVM:
www.fda.gov/cvm/

For having wool tested in a laboratory (Yocom-McColl Labs): www.ymccoll.com/
For a list of small-scale wool processors: www.rbparker.com/Processors.html
See also www.rbparker.com/ for some other information of interest to sheep raisers.

Disclaimer: email addresses and Web URLs such as those listed in this appendix and elsewhere in *The Sheep Book* often change unpredictably. If one fails to work, try searching using a search engine such as www.google.com.

INDEX